ISCS SandFest 2013

ISCS

November 21, 22, 23, 24
Tellus Science Museum
Cartersville, Georgia

Edited by Lori Carter

A completely volunteer, non-profit 501(c)(3) organization dedicated to educating the youth of the state and members of our society in the field of Earth Sciences

All proceeds above cost will be used to further education in Earth Sciences and to support educational efforts of the International Sand Collectors Society.

This edition published by:

The Georgia Mineral Society, Inc.
4138 Steve Reynolds Boulevard
Norcross, GA 30093-3059
www.gamineral.org

ISBN: 978-1-937617-02-8

Cover:
California Dunes, Aruba by Leo Kenney
Bookworm Drawing by Lori Carter

Table of Contents

POETRY

SAND ART

HUMOROUS ART

Julian Gray

Increasing Depth of Field in Sand Photography

Equipment

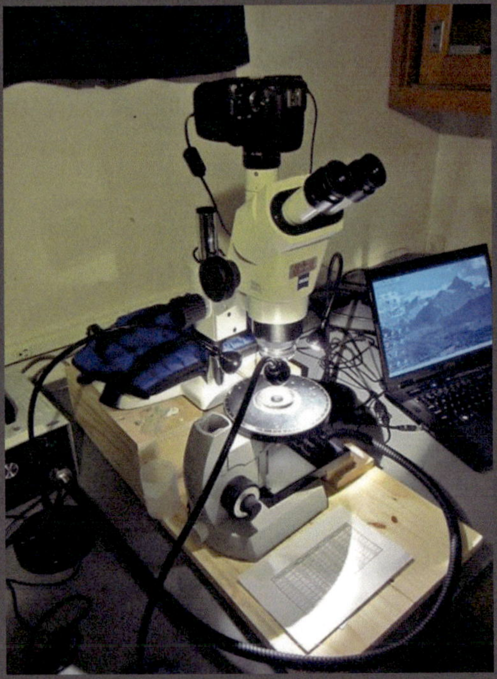

- Canon T3i Digital SLR
- Zeiss Stemi 2000-C
- Sample
- LED Desk lights
- Fine Focus Stage

Sample Handling Technique

The Problem:

- As magnification increases:
  - Resolution increases
  - Depth of Field Decreases

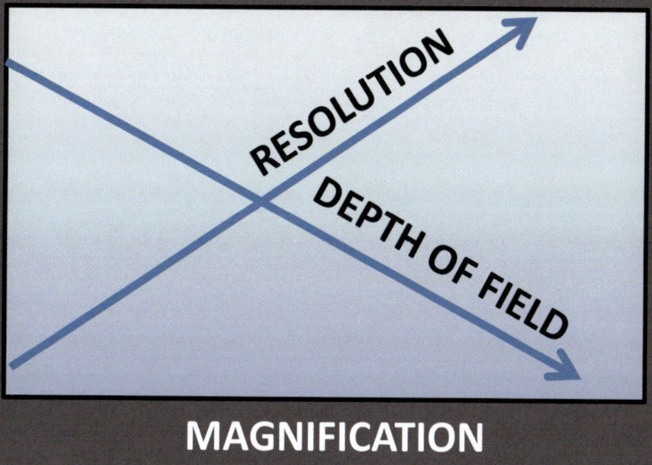

The Solution:

- Focal Stacking:
  - Take multiple image at different focus depths
  - Process with software that retains sharpest part of each image
  - Create sharp composite image

Stacking Software

- Combine ZP (freeware)
- Helicon Focus
- Adobe Photoshop
- Zerene Stacker

Focal Stacking: The Finer Details

- Depth of Field must be known, calculated, or measured
- Consecutive images must overlap
- Rule of Thumb:
  - Focus steps of ½ Depth of Field
  - Ensures each part of subject is sharply recorded three times
- Focus movements must be controlled precisely

Measuring Depth of Field

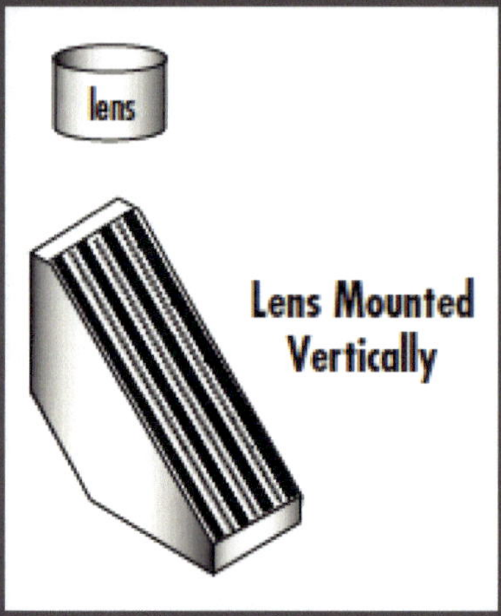

Depth of Field Test Target offered by Edmunds Optical
(DoF Target 5-15, #54-440, $575)

Measuring Depth of Field

- Economical homemade Depth of Field Setup:
  - Precision Stage Micrometer at 45 degrees
  - Requires simple calculations

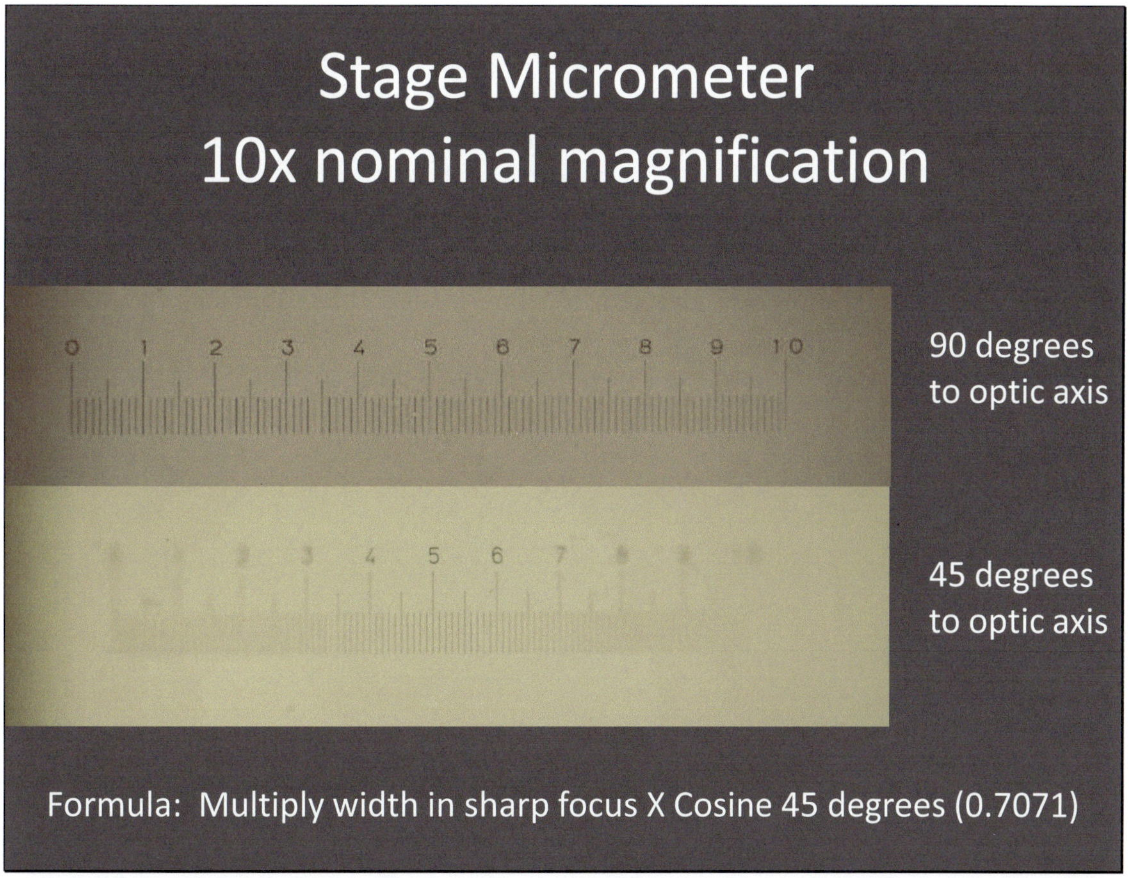
Stage Micrometer
10x nominal magnification
90 degrees
to optic axis
45 degrees
to optic axis
Formula: Multiply width in sharp focus X Cosine 45 degrees (0.7071)

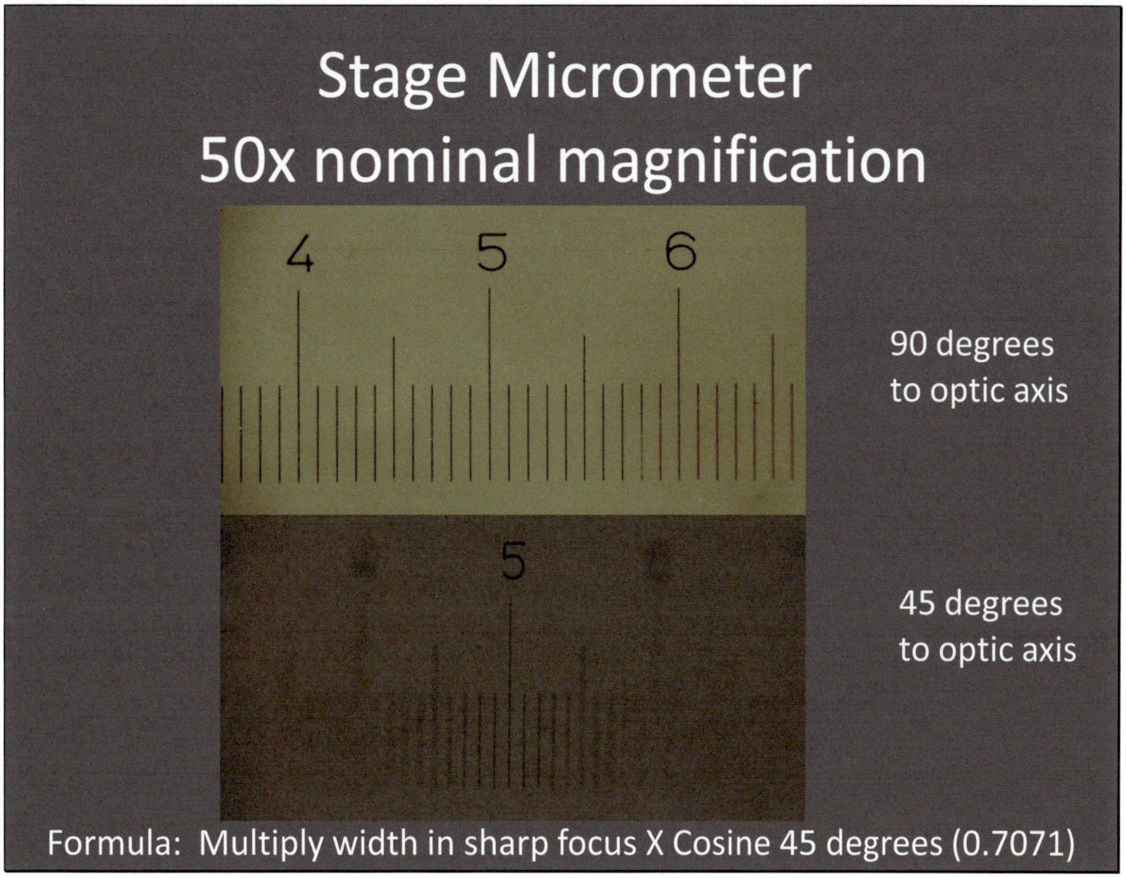
Stage Micrometer
50x nominal magnification
90 degrees
to optic axis
45 degrees
to optic axis
Formula: Multiply width in sharp focus X Cosine 45 degrees (0.7071)

Sample Depth of Field Calculations

Nominal Objective Magnification	X (mm)	Z = DoF (mm)	Z = DoF (μm)	0.5 DOF (μm)
5.0	0.1	0.0707	71	35
4.0	0.15	0.1061	106	53
3.2	0.2	0.1414	141	71
2.5	0.25	0.1768	177	88
2.0	0.3	0.2121	212	106
1.6	0.4	0.2828	283	141
1.25	0.7	0.4950	495	247
1.0	0.8	0.5657	566	283
0.8	0.9	0.6364	636	318
0.65	1.5	1.0607	1061	530

Controlling vertical movement

- Z-axis positioner with micrometer
- Motorized stage with stepping motor
- Fine focus knob

Coarse focus only. Poor Precision, no control

Z-axis positioner with micrometer

Motorized stage with stepping motor

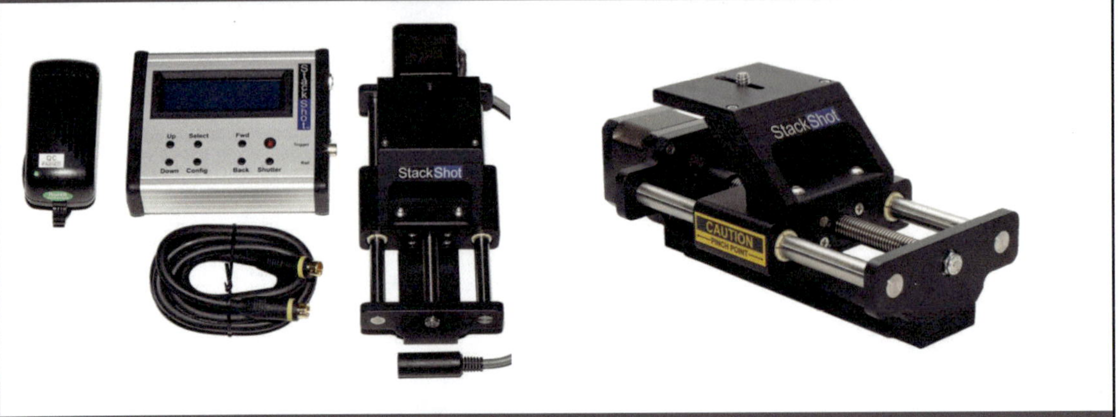

- Stack Shot Macro Focus Rail by Cognisys, Inc.
- Programmable
- Precise movements as small as 2 microns

Fine Focus Knob

- Zeiss Standard 16:
  5 micron steps

- Leitz Ortholux I:
  1 micron steps

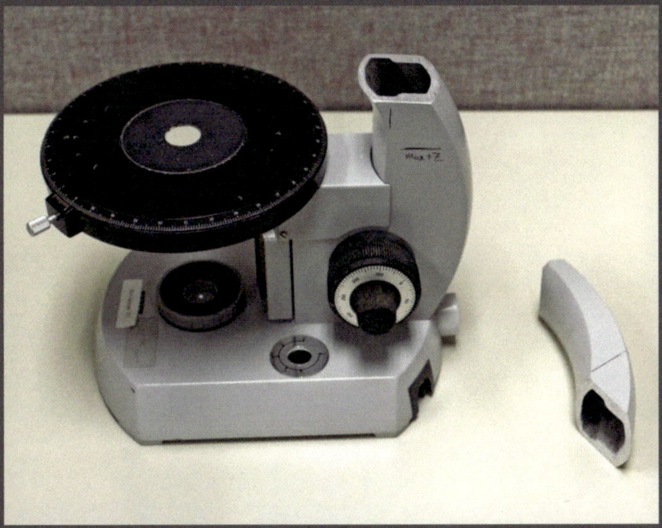

Used microscope

Arm removed to fit under standard stereomicroscope

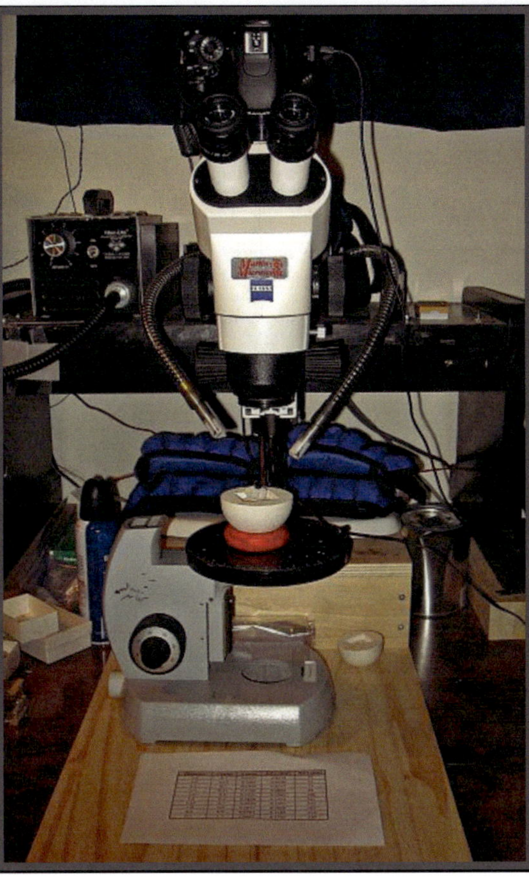

A setup using fiber optic illumination

Further Refinements on Photography: Diffusion

Diffusion Acetate ring or tent to surround sample

Sand being photographed using a diffusion tent

Putting it all together

Results and Examples

No Diffusion

With Diffusion

Heavy Mineral Concentrate Sand
Northern Florida

River Sand
Stewart County, Georgia

Oolitic Sand
Great Salt Lake, Utah

Star Sand
Taketomi Island, Okinawa, Japan

Star Sand
Taketomi Island, Okinawa, Japan

Olivine Sand
Main Island, Hawaii

Olivine Sand
Main Island, Hawaii

Olivine Sand
Main Island, Hawaii

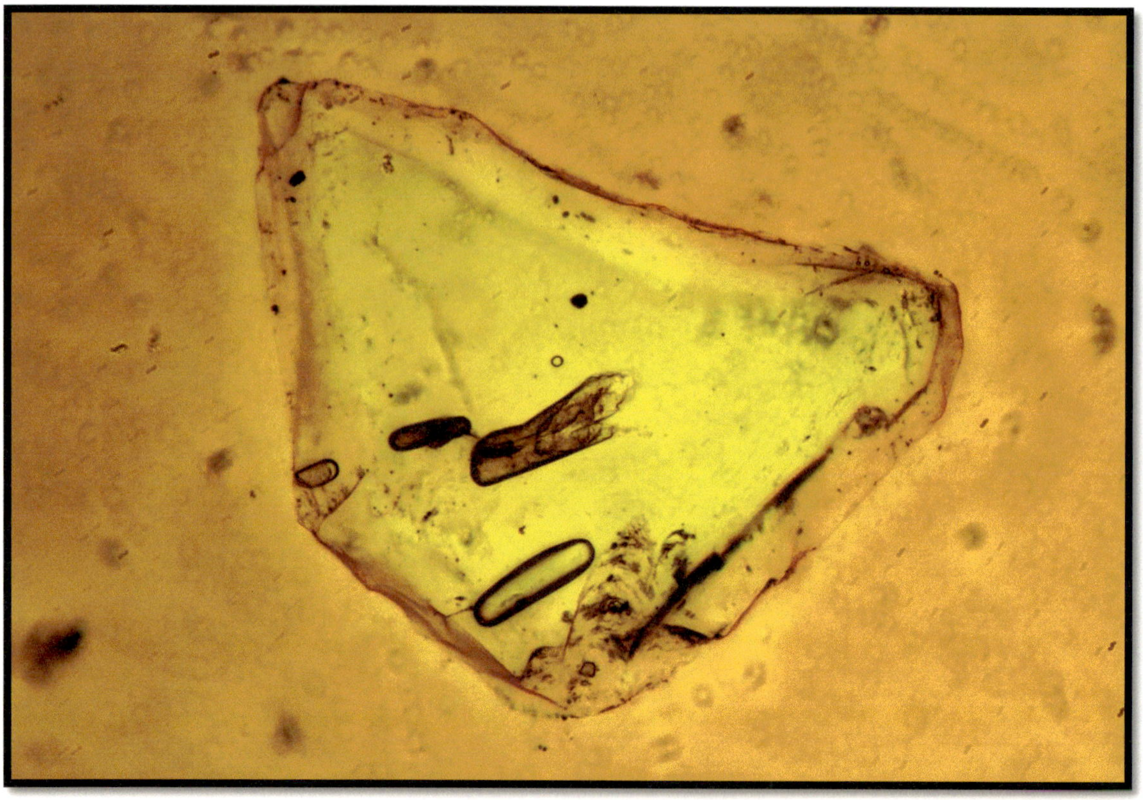

Garnet Grain with Inclusions

Ruby Beach Sand
Ruby Beach, Olympic Peninsula, Washington

Conclusions

- Focal stacking is easy to use and produces quality photomicrographs of sand
- Diffusion can be used judiciously depending on reflectivity of subject matter

Leo Kenney

Basic Sand Photography

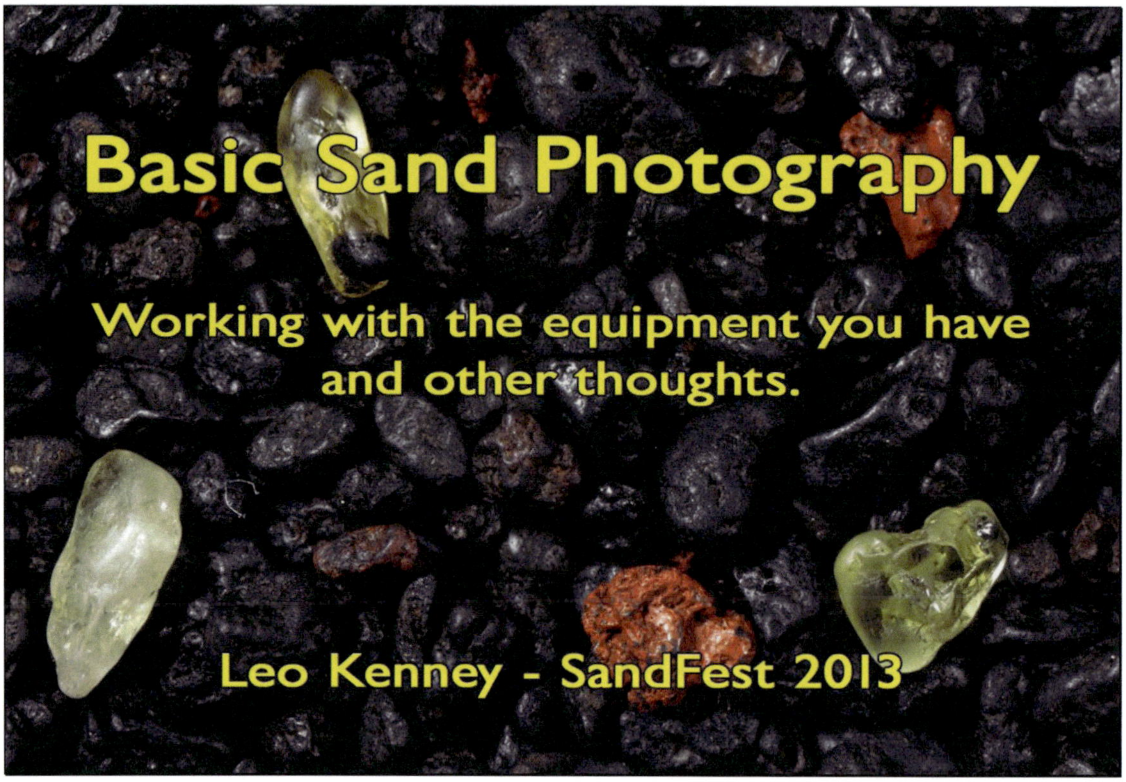

For photographing sand, you have three levels of camera equipment available: the point-and-shoot camera which pretty much makes all choices for you, the very popular all around fixed lens adjustable digital camera which allows considerable photographer input, and the digital camera with interchangeable lens for which you could use a dedicated 1:1 (or better) macro lens. The intended use of your photograph as well as your skill, time commitment and budget are all involved when you ask "Can I take close-up pictures of sand?" For simple uses, web page or small print, any of the equipment will get you some level of acceptable photograph. For enlargements, great detail, scientific work or publishing, then a macro lens and other equipment is the name of the game. However, as with Goldilocks, for most people the middle choice might just be good enough.

Since this presentation is "basic", I will skip a lot of photo jargon and many details. If you want further information, please talk to me. For me, the reality is I cannot teach sand photography in 20 minutes when one considers all the variables of equipment, technique, and technical level of the photographer.

Better yet, I will list several internet resources which are very useful for both digital photography and sand photography.

For the simplest of close-up photographs, steady your camera with a tripod, have plenty of light on the subject, set the camera to “macro”, focus, and fire away. To further decrease camera movement, use the self-timer rather than pressing the release button, or use a remote release.

Your camera is determining how much light it senses. If you can choose where to measure light, then try a number of areas, both lighter and moderate in brightness. The camera will determine lens opening and speed. Take a few pictures.

This is what you get!

I suggest including a scale and possibly a white card and a gray card in at least one of the images. The scale will help to determine magnification. The white/gray cards will be useful for adjusting white balance and levels later. The cards are also useful for determining light levels when photographing the sand.

The best way to keep track of the scale for your sand image is to make a Photoshop layer with the scale on it. This could be either a copy of the scale itself or the scale bar you make. Move that layer somewhere into the area you will crop for your photo. Here the scale bar is visible but it could be on a hidden layer. Crop the image so that it pleases you. Many lens distort the outer edges of a macro photo so you might want to crop a bit to eliminate that distortion when cropping. The lighter area is the area of the crop.

Now you have a sand image with a scale bar. Move the bar as you wish or hide it. As this point, for the most basic image enhancement, I would use Photoshop to adjust levels. For most cases, “auto levels” will work.

Your completed sand image! I cropped the original to 6” x 4” at 200 dpi which would provide a reasonable 6 x 4 print or an image for a web page or even powerpoint presentation (I hope). There are more “things” you could do with Photoshop but you won’t hear about that here.

When I agreed to this presentation, I asked Lori Carter to take a few pics using her camera using her current technique. I was curious to see the results with setting the aperture (f-stop or lens opening) to different values. A smaller lens opening (high f-stop number) gives a greater depth of field. Depth of field is how much, in depth, of the image is in focus. In close-up photography, we usually want as much depth of field as possible. (Some people also compensate for poor depth of field by flattening the sand in the dish.)

Once you get out of auto mode with your camera, you need to consider speed, aperture, and ISO as all three need to be balanced for a perfect exposure. In "manual" setting you control all three. There are also semi-auto settings in which you might choose aperture or speed and the camera chooses the rest.

ISO is a setting which determines how sensitive the camera is to light. A high ISO (say 800) means the camera can produce an image with less light than a lower is ISO of 100. Increasing the ISO introduces more noise into the image.

A smaller aperture (higher f-stop value) gives greater depth of field yet requires more light.

In this test, the aperture was set and the semi-auto settings of the camera determined speed and ISO.

For a discussion of aperture, speed and ISO try this blog which has excellent instructions for understand digital photography.

http://digital-photography-school.com/

There is no dramatic difference between the test shots; both would be usable sand images. However, the f/8 was expected to be the sharper of the two, but then, the camera changed the ISO setting so we have two variables. Also, auto-focus was used for these shots. Or maybe there is no real difference in lens settings in the macro mode with this camera.

I know you can get this far with most digital cameras. You should try mutliple lens settings, take lots of pictures, learn some Photoshop skills, and have fun. Research and practice will improve your photographs. If they never meet your expectations, better equipment might improve the results but only with further research and practice.

We will switch to my equipment for a few more points.

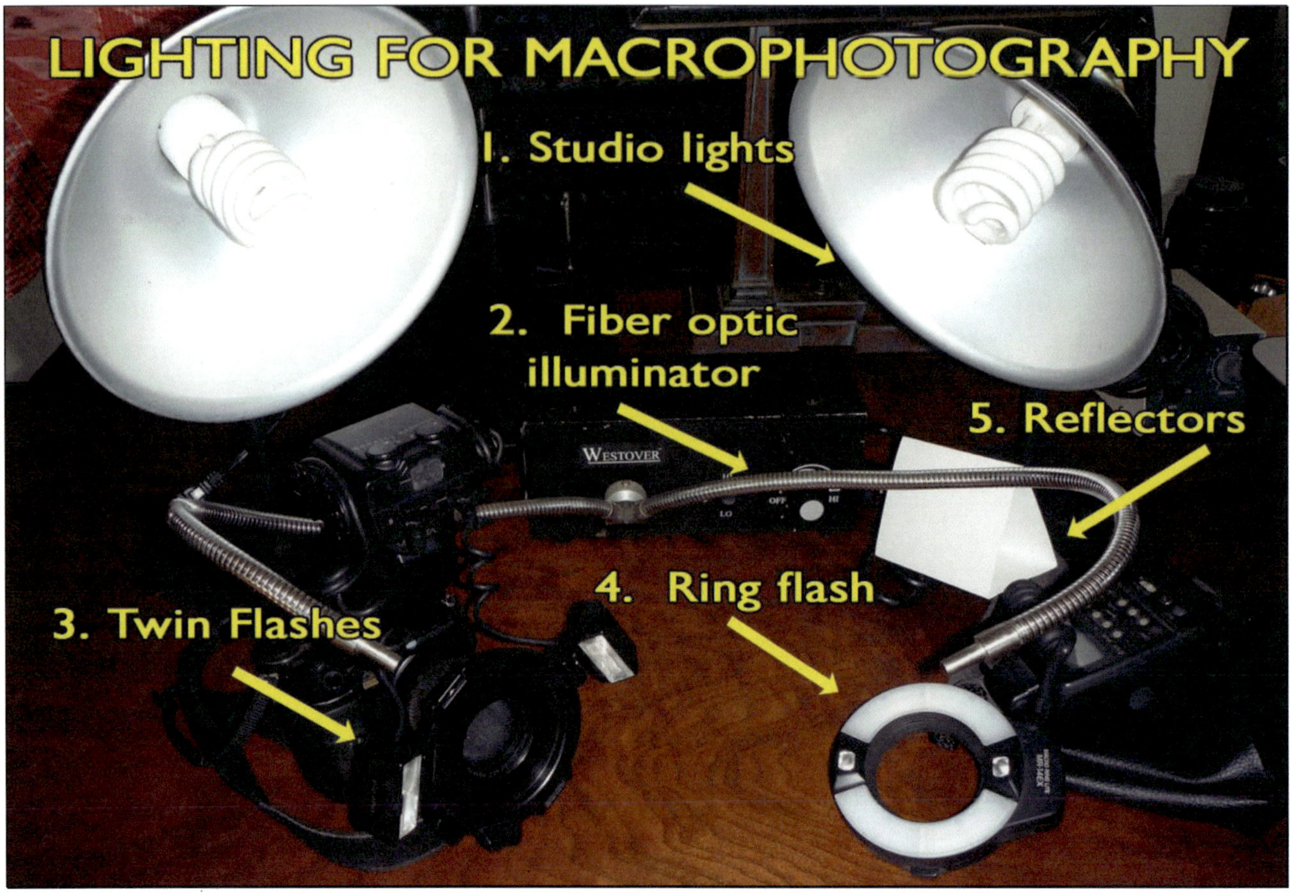

As you try to use small aperture, low ISO, and the fastest possible speed, you will need a good source of light. Sunlight works well but has limitations.

Your on-camera flash might provide light but it will be from directly above and not have natural shading to give depth to the sand. Here we have representation of some possible light sources.

- Studio lights with lamps that mimic the color temperature of sunlight.
- Fiber optic illuminator which allows you to carefully direct two beams of light.
- A pair of flashes, either on the camera as shown here, or off camera.
- A ring flash that attaches to the lens. This model has a right and left unit so is actually two flashes.
- All photographers use reflectors to direct light where it is needed and/or to produce special effects.

Whatever your light source, aim for an exposure that is properly exposed. An overexposed image (left) noticeably has the lighter grains washed out and with no detail. The right image is a little dark but it has good color and detail. We can work with the right hand image in Photoshop.

The previous darker image after using auto-levels in Photoshop.

I always take at least 3 shots at different exposures. Here are thumbnails for three sands.

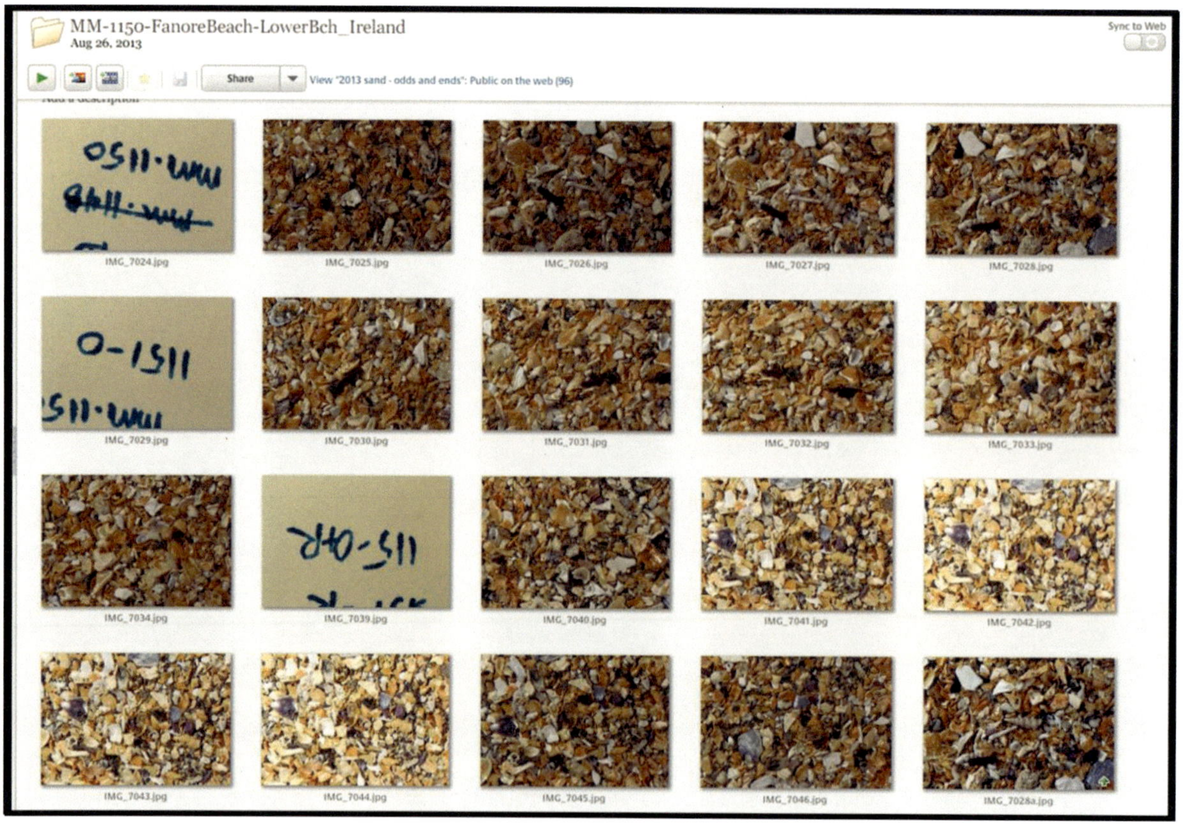

In some cases, I take many shots looking for good color and lighting and working on composition.

Let me introduce you to my current equipment for sand photography.

This is the studio. To the left is an adjustable digital camera used for “actual size” images of sand. In the center is a microscope, often used to arrange sand grains. It is on a swivel arm so it can be pushed out of the way. To the right is the Canon EOS system I use for sand. For the curious, the background equipment is a large format printer.

1. Canon makes some fine digital cameras. This is the EOS 60D with Canon's 65 mm 1-5x macro lens. This lens is only for macro work. Other lens have a 1X macro function in addition to general use. 1-X (or 1:1) macro means the the field of view you see when you compose your image is equal to that of a 35mm slide. Since a digital sensor is smaller, magnification is actually greater than 1:1 with a digital camera. At 5-x, a grain of sand will almost fill the screen but taking that type of image presents its own problems.

2. Twin flashes provide lots of regulated light to allow for short exposures with small f-stops as well as other advantages. These flashes produce the correct light for the manual settings chosen on the camera for the image.

3. If one wants to set the lens for a certain magnification such as 1-x, then the actual manual focusing is done by moving the camera to and from the subject. An adjustable copy stand is a real luxury, more likely a necessity, if one wants to shoot consistently at a set magnification. Focusing rails and other add-on equipment also work for moving the camera in small amounts for focusing.

4. The fiber optic illuminator throws plenty of light so one can see the sand to focus. Some people use this as the light source for their photography.

Not obvious in this illustration is a remote release so one does not jiggle the camera during a photograph.

Siim Sepp on his blog (http://www.sandatlas.org/) presents a nice explanation of sand macro-photography with some of this equipment.

Just to show the capabilities of a 1x macro lens, here is the same foram sand from Japan taken with the Canon lens at 1X with the previous photograph as an insert. Note the difference in sharpness and detail.

Even with only 1x magnification and with 18 megapixel resolution, the image is capable of quite a bit of enlargement and is suitable for large prints and professional detail work.

This is a crop of an image taken at 3x with the Canon 65 mm 1-5x macro lens.

So what does one do with pictures of sand?

Sand photography is the confluence of science and art. I like my photographs to be true to the collected sample and, if possible, attractive. My goal is maintain that scientific integrity and make the image visually interesting. This sample has nice content and composition and tells an interesting story. The grains have been sorted a bit and arranged using fine forceps. The more interesting grains are the three larger pieces of amber (fossilized tree sap) and the encrusted black oil blob, probably from one of the off-shore oil platforms.

Some collectors use sand photographs for their galleries, trade lists, blogs, etc. Steve Benton is one of many collectors with some of their sand on line. I've been helping Steve by photographing some of his extensive collection.

You can see some of Steve's sand here:

https://picasaweb.google.com/104622853190354609839/

or

https://plus.google.com/photos/104622853190354609839/

Email: NolzRule23@aol.com

Connecting the sand image with the sand source is educational and fun. Many photos of beaches are available on the internet and may be used under Fair Use. It is also possible, as mentioned earlier, to take a reasonable close-up of sand at the site and then photograph the site at the same time.

I like to make prints, posters, calendars and other art. This poster is a thank-you to the person who sent me a tremendous amount of Alaskan sand.

Iceland. Actual size is 22"x11"

In this case, an individual agreed to collect sand while in Iceland if I would put it into a form she could display.

As a retired teacher, I like to encourage science in the classroom. The students of a science teacher at the Thurston Middle School, Westwood, MA collect sand as an exercise in labeling with a little geography thrown in. I end up with the sand and produce both an 8' X 10" page for each student and an image for the teacher's website. Bonus, I get to keep the sand. Some of the Thurston sand galleries are here: https://picasaweb.google.com/108544250519116867331

Kurt Meyer is a collector/photographer with a website "Tropics of Sand". He is also producing prints of his images. Kurt's website is here: http://www.tropicsofsand.com/

Kurt uses a 1:1 60 mm Nikon lens on a Nikon D7000 body which takes a16.2 megapixel image. His images are artistic with interesting arrangements of the grains. He uses a lot of black space to highlight the individual grains.

Kurt Meyer wrote a guest blog for Michael Welland's "Through the Sandglass" blog. Kurt discusses the styles of different sand photographers.

http://throughthesandglass.typepad.com/through_the_sandglass/2012/04/guest-post-a-gallery-of-sand-photographers-styles.html

Otama Beach, Waikato, NZ by Carla Lagendijk

My favorite sand collector/photographer/artist is Carla Lagendijk. She produces remarkable photographs with a Nikon Coolpix 4500 (four megapixel sensor) using a tripod, sunlight, and Photoshop. Her images are suitable for most uses not requiring a large print. They are certainly suitable for her uses as you will see. "My goal is not to cover a whole wall with a picture."

Rochebonne Beach, St. Malo, France by Carla Lagendijk

Yes, she does images of entire sand samples and does them well.

selection - Enubuj Island, Marshall Islands
by Carla Lagendijk

But the real treat, would be the "selections" she makes where individual grains are arranged in an interesting pattern and photographed.

selection - fossil forams, Chattien, France
by Carla Lagendijk

I love to look at these. I have not tried this style myself.

Carla Lagendijk's educational displays are also amazing and show more uses for sand photographs. These are two of a series of large shadow frames done for educational displays in her home country of the Netherlands. Note the use of location images, sand images, and actual sand samples.

I like the use of the micro-boxes of sand on top of the image of the sand.

A closer view of the previous frame.

Peter Blau

Light Microscopy of Sand: Lighting, Scaling and Image Enhancement

A spectrum of collecting

Geologic formations → Landscaping boulders → Museum-sized → Hand-sized → Micromounts → Thumbnails → Sand

A lucky rockhound

Sand collecting takes you to some interesting places

Technically, the definition of 'sand' is based on size, not composition

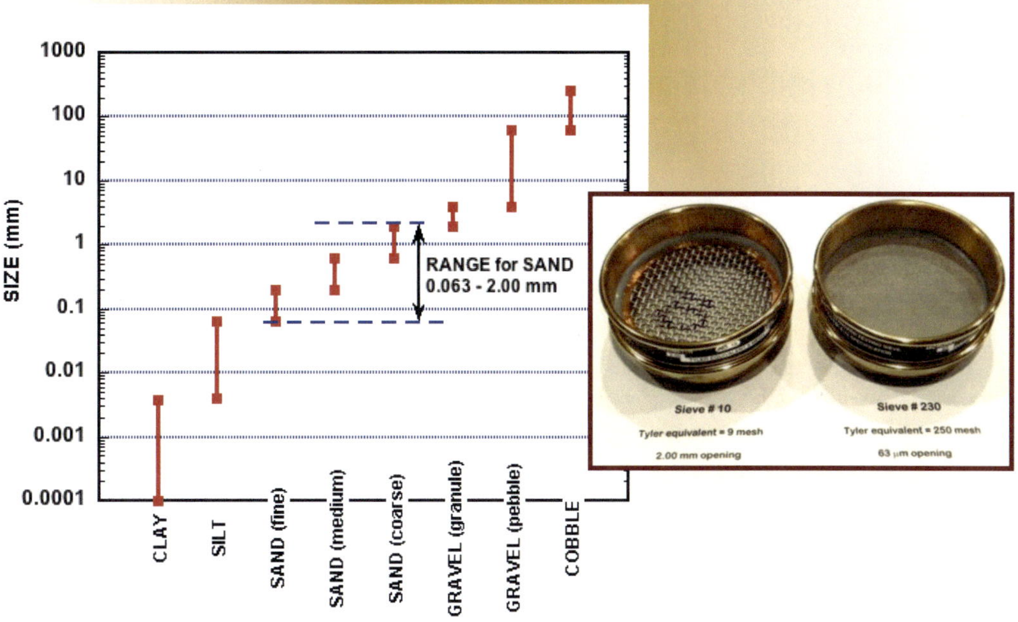

Sand collecting, the hobby

It's a 'vial' hobby

- 3 dram flint glass vials
- 16 mm dia. x 55 mm long

How much is enough?

Description	Mean grain diameter (mm)	Number of grains in the vial*
Coarse sand	2.0	1,526
Medium sand	0.42	164,800
Very fine sand	0.063	50,000,000

* spherical grains with a packing factor of 0.60

Tools of the arenophile

SAND

Outline

- Sand: collecting in a world of micro-minerals
- Image acquisition, adding scaling
- Transmitted, reflected light, and combinations
- Cross-sectioning and etching sand particles
- Meteoritic sand and lunar soil
- Image gallery

Computer / display

Digital camera

Projector lens

Twin oculars

Zoom ring

Objective nosepiece

Illuminator light pipes

Specimen and stage

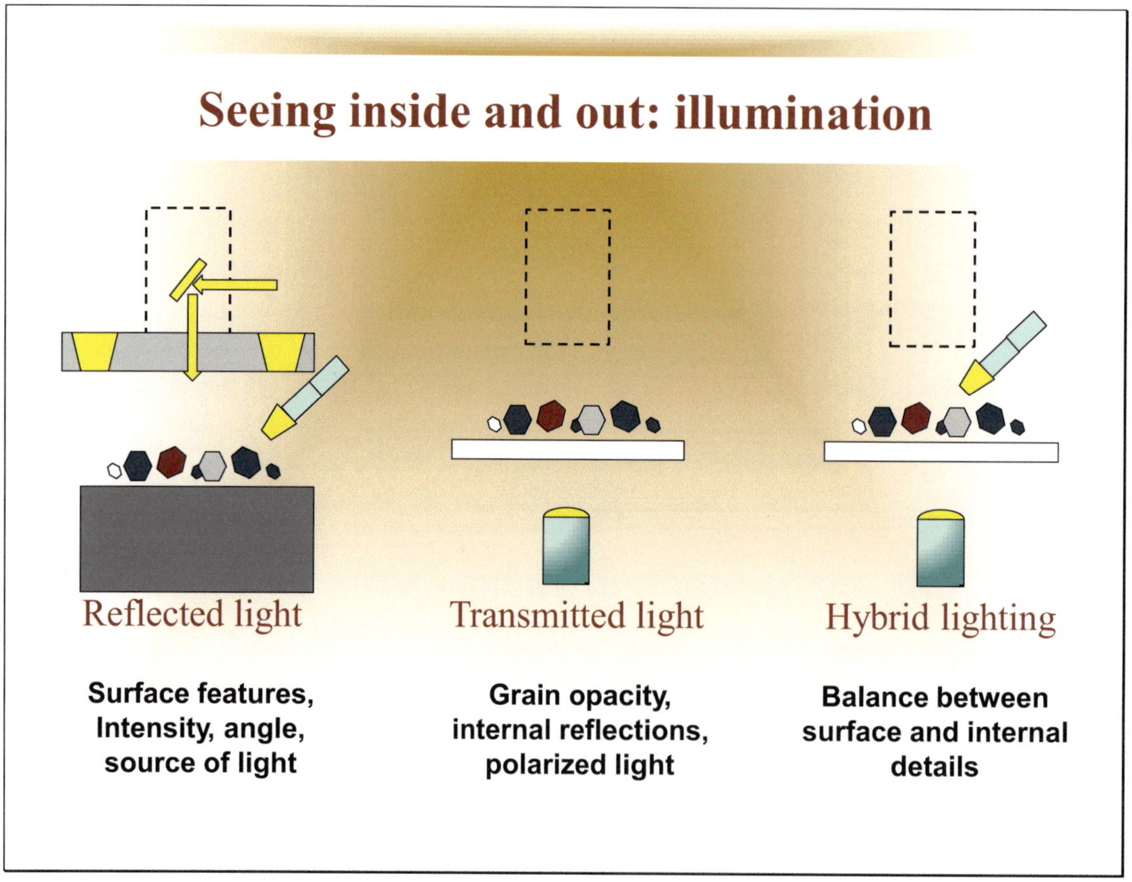
Seeing inside and out: illumination
Reflected light
Transmitted light
Hybrid lighting
Surface features, Intensity, angle, source of light
Grain opacity, internal reflections, polarized light
Balance between surface and internal details

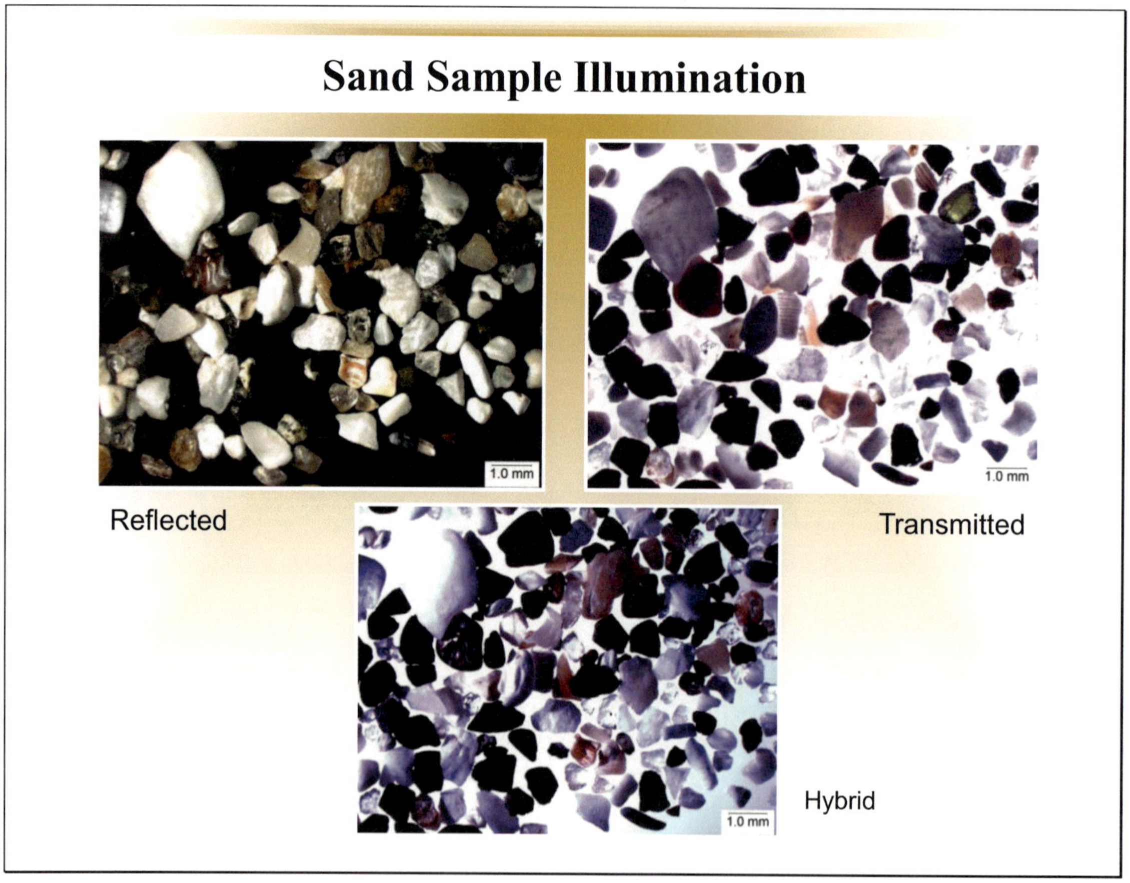
Sand Sample Illumination
1.0 mm
1.0 mm
Reflected
Transmitted
1.0 mm
Hybrid

Through the microscope -

Polarized light reveals the interior of the singing grains

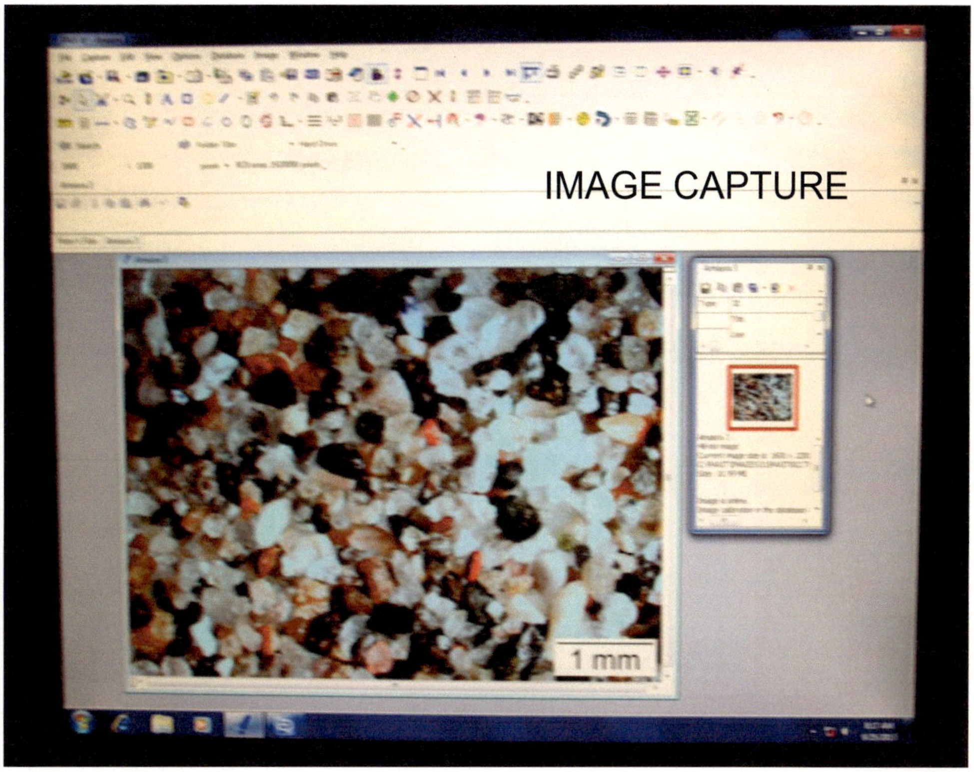

Some imaging demons: washout, over saturation, blurring, contrast and shadow detail, spurious light reflections, true-color (sunlight color), depth of field limitations.

Magnification on a page is not the same as on the movie screen!

Obtain a high-precision ‘stage micrometer’

Focus on the stage micrometer at all magnifications you intend to use.

Imaging software allows you to select a reference length and superimpose it on the scale bar. This sets the calibration for each magnifying combination.

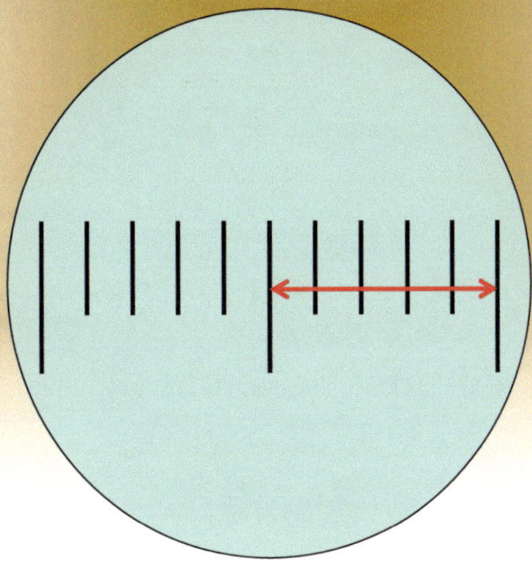

Bar appears on the screen. Stretch to fit the scale, then tell the software low long the bar is supposed to be

MENU BARS OFFER OPTIONS

Storing, naming, measuring features, labeling features, adjusting contrast, brightness, coloration, sharpness…

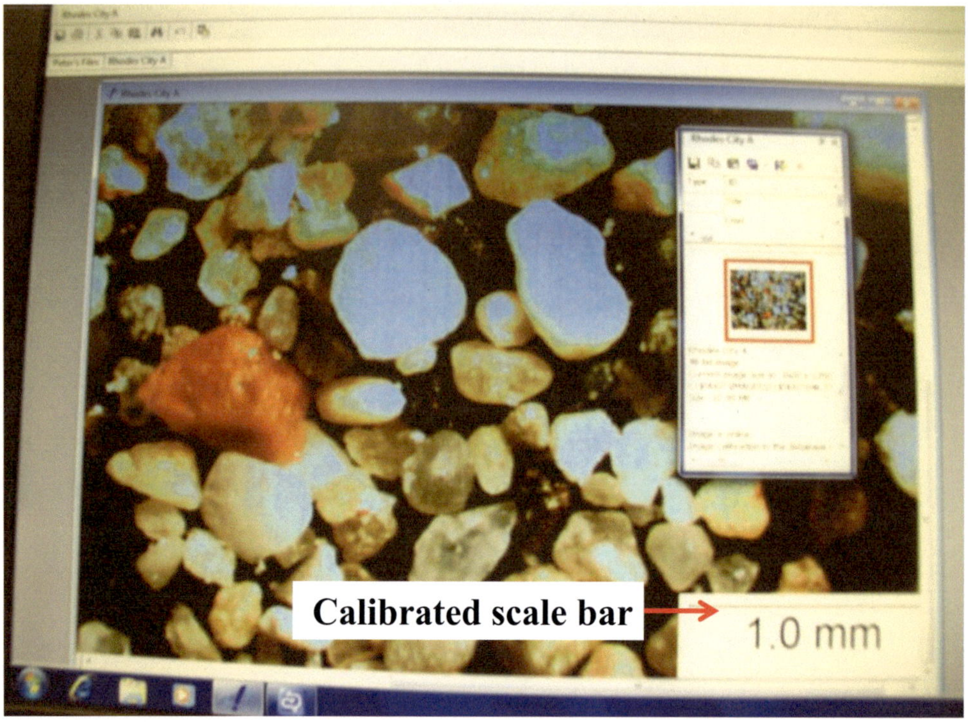

Through the microscope (1)

Mt. Etna, Sicily
(August 2011 - Z. Bell)

Aqtau, Kazakhstan
(2010 - Z. Bell)

Through the microscope (2)

White Sands, NM
(2010 - R. Hohenbrink)

Herring Cove Beach, Cape Cod MA,
(2011 - B. Beckerman)

Through the microscope (3)

Bottom of the Grand Canyon, Arizona
(2010 - D. Blau)

Shekna River, Russia
(2008 - Lukin-Beck)

Through the microscope (4)

Waikiki, Hawaii
(2008 - S. Miller)

Punaluu, Hawaii
(2008 - S. Miller)

Hawaiian Sand Jewelry by Michael Cohen

- Oahu: Lanikai Beach
- Oahu: Sunset Beach
- Maui: Kamaole Beach
- Kauai: Kalalau
- Oahu: Kahaluu
- Hawaii: Hilo Bay

Through the microscope (5)

Sail Hotel Beach, Dubai UES
(2010 - J. Treadwell)

Weasel Cave, N. Ossetia, Russia*
(2011 - L. Kimbal)

* Ash from a 200K YBP volcanic eruption. Neaderthal dig site.

Through the microscope (6)

Godafoss Falls, Iceland
(Aug 2006)

Geysir, Iceland
(August 2006)

Through the microscope (7)

Lawrencetown Beach, Nova Scotia
(July 2011 - A. Payzant)

Glacier Bay Point, Alaska
(August 2006 - T. Besmann)

Through the microscope (8)

Dunes of Sossuvlai, Namibia
(August 2007 - F. Vandervoort)

Truckee River, NV
(August 2006 - F. Vandervoort)

Through the microscope (9)

Peñiscola, Spain
(October 2011 - Z. Bell)

Valencia, Spain
(October 2011 - Z. Bell)

Through the microscope (10)

Island of Okinawa
(October 2011 - Z. Bell)

Sail Hotel Beach, Dubai
(March 2010 - J. Treadwell)

50,000 years, ago a meteor impact created fine particles of metallic sand ('metallic spherules') near Canyon Diablo, Arizona.

Cross-sectioning of Canyon Diablo and lunar soil particles

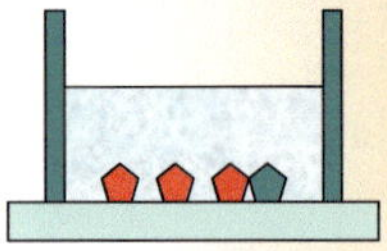

1) Place mold release on a flat glass plate
2) Sprinkle and position particles
3) Place ring mold
4) Mix slow-set epoxy and let the bubbles rise
5) Pour ~ mm very carefully
6) Allow bubbles to rise
7) Fill the rest of the mold
8) Cure overnight and carefully slide off the glass
9) Use abrade by hand to flatten the face
10) Polish with a series of diamond abrasives
11) Etch with 2-5% nitric acid in alcohol
12) Examine in the light optical microscope/SEM

Cross-sections of Fe-Ni-Co-S-P spherules: Simulates those from impact-melted meteors (mounted, polished, and etched with 2% Nital.*

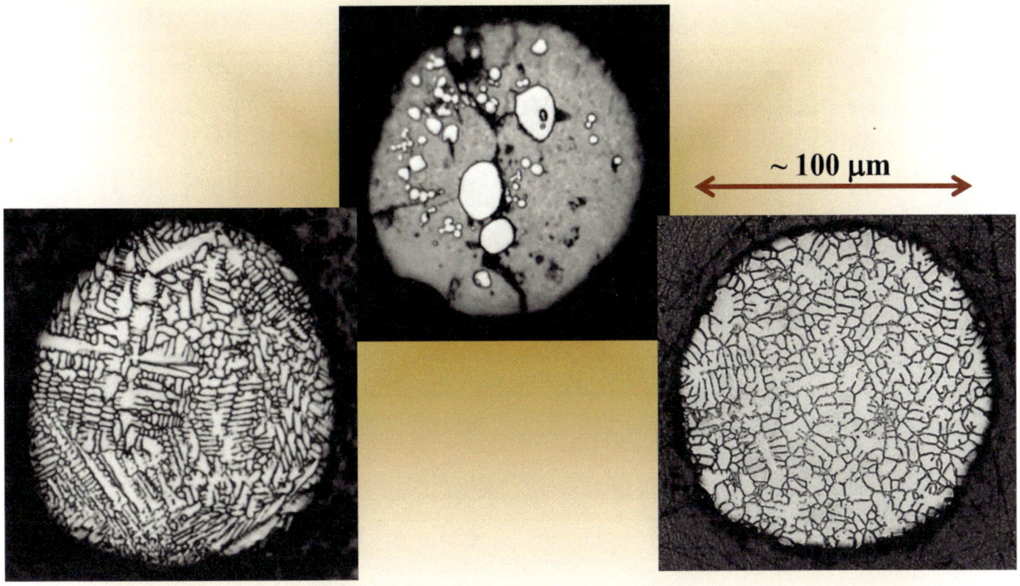

* Ref:. P. J. Blau and J. I. Goldstein (1975) "Investigation and Simulation of Metallic Particles from Lunar Soils," *Geochim. et Cosmochim. Acta*, Vol. **39**, p. 305.

Advanced techniques are also available for research and forensic studies

- Confocal microscopy
- Light section microscopy and image stitching
- 3D microscopy – multiple oblique light sources (see Dr. Gary Greenberg (2008) "A Grain of Sand," Voyageur Press)

Summary

◆The hobby of sand collecting is enhanced by using light microscopy to examine, document, and present your collection.

◆Illumination control is the key to bringing out the most interesting features of the samples.

◆Mounting methods from loose examination to cross-sectioning were described.

◆Software can reveal features and allow scale bars to be inserted.

Sand is a measurer of time

Thank you and good bye!

Thom Hopen

PLM Analysis of Mineral Grains

Polarized Light

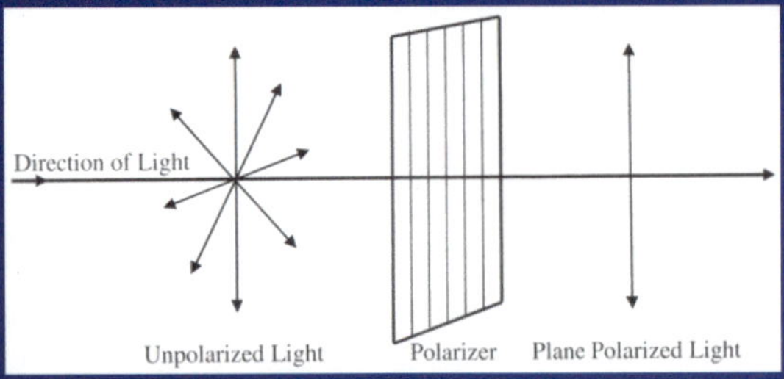

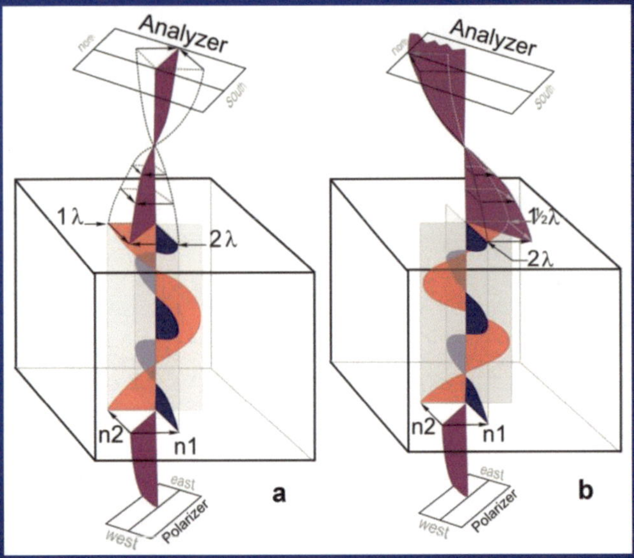

Polarized Light Microscope

Sieving (wet or dry

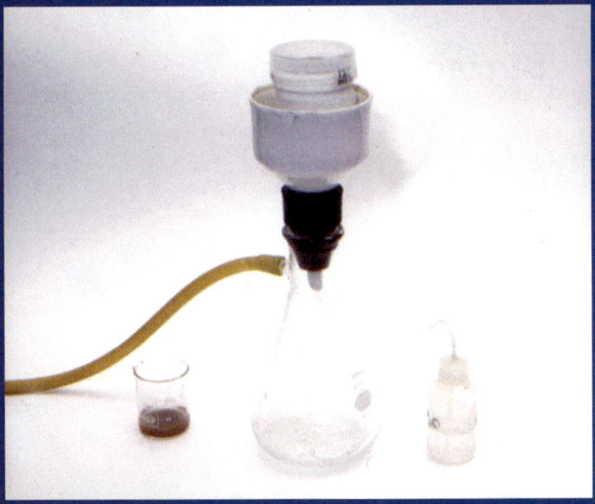

Density Separation

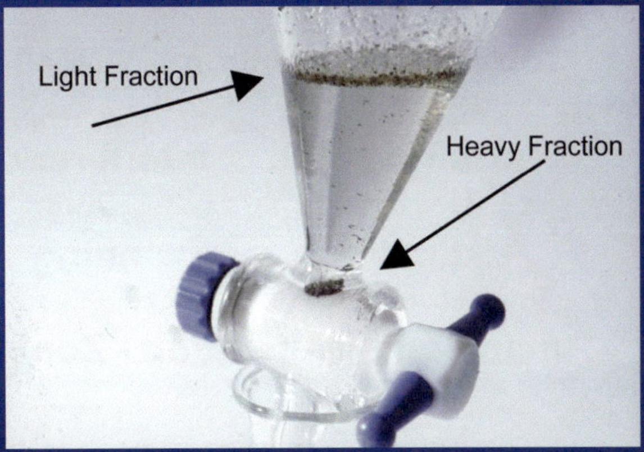

Specific Gravity of Common Minerals

Float on Bromoform

Quartz (2.65) Feldspars (2.5-2.8)
Calcite (2.7) Gypsum (2.3)

May Sink in Bromoform

Micas (2.7-3.4) Dolomite (2.8-2.9)

Sink in Bromoform

Amphiboles (2.9-3.5)	Limonite (3.8)
Apatite (3.2)	Olivine (3.3-3.4)
Chlorite (2.6-3.0)	Pyroxenes (3.3-3.6)
Corundum ((3.9-4.1)	Rutile (4.2-4.3)
Epidote (3.2-3.5)	Sillimanite (3.2)
Garnet (3.8)	Straurolite (3.6-3.8)
Hematite (5.2)	Tourmaline (2.9-3.2)
Kyanite (3.6)	Zircon (4.7)

Mineral Characterization by PLM

Size Morphology Surface Texture

Crystal System - Refractive Index (indices)

Color – Pleochroism Relief – Dispersion Staining

Birefringence Interference Figure Optic Sign

Extinction Twinning Sign of Elongation

Inclusions Alteration - Zoning Microchemical Test(s)

Grain Characterization

Morphology

Crystal Form
Habit or Shape
Sphericity
Angularity or Roundness
Euhedral
Subhederal
Anhedral
Cleaved Grains
Fractured Grains
Broken Grains

Grain Characterization
Crystal Form

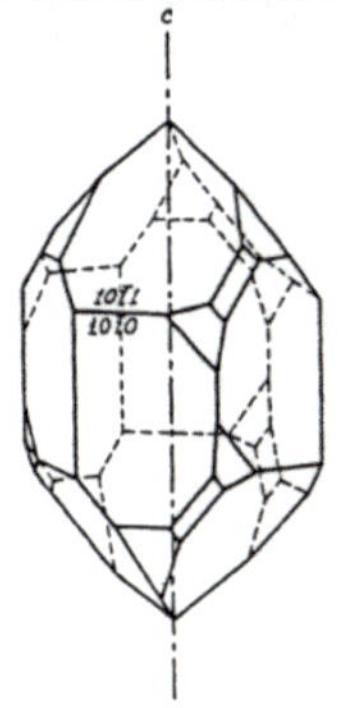

α-Quartz

$n_\omega = 1.544$
$n_\epsilon = 1.553$
$n_\epsilon - n_\omega = 0.009$
Uniaxial positive

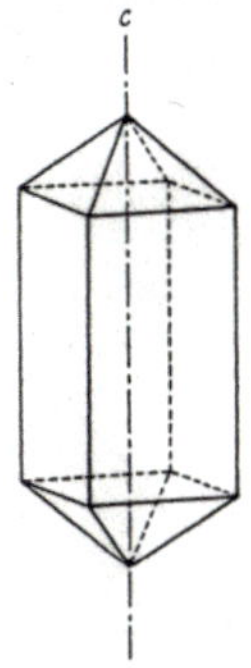

Zircon

$n_\omega = 1.920–1.960$*
$n_\epsilon = 1.967–2.015$
$n_\epsilon - n_\omega = 0.042–0.065$
Uniaxial positive (rarely biaxial positive $2V_z < 10°$)

Grain Descriptions
Shape or Habit

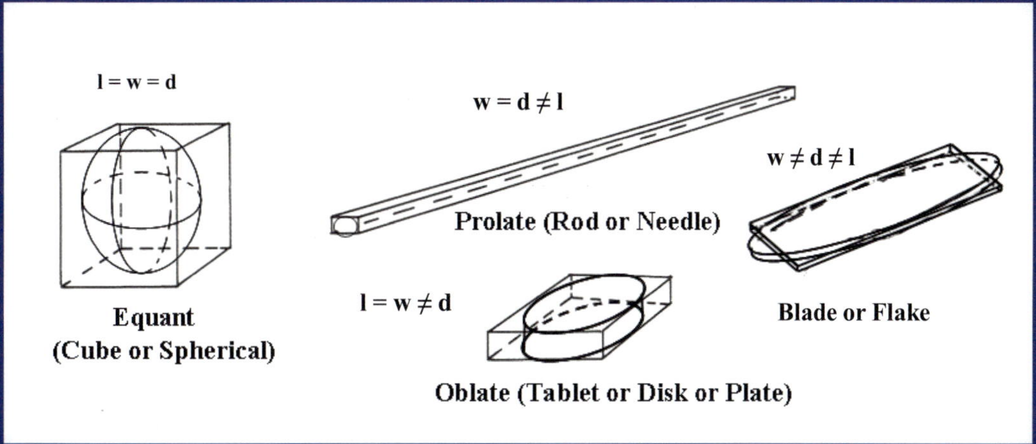

Grain Descriptions

Angularity - Sphericity

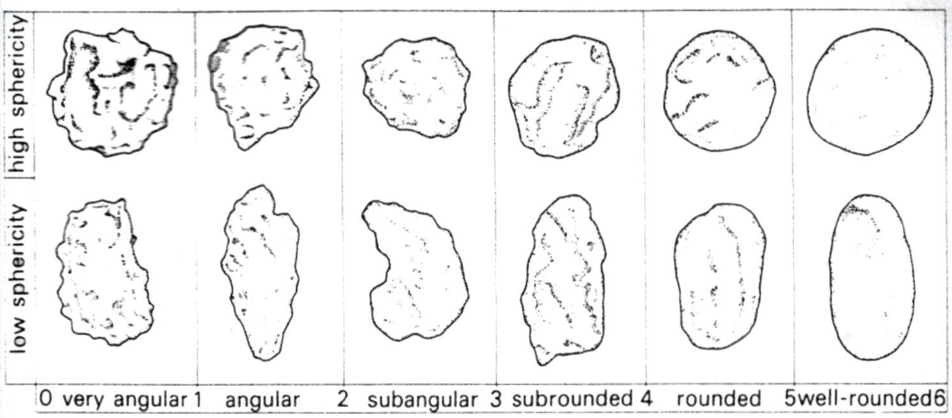

Categories of roundness for sediment grains. For each category a grain of low and high sphericity is shown (after Pettijohn *et al.*, 1973).

Subhedral

Zircon Grains

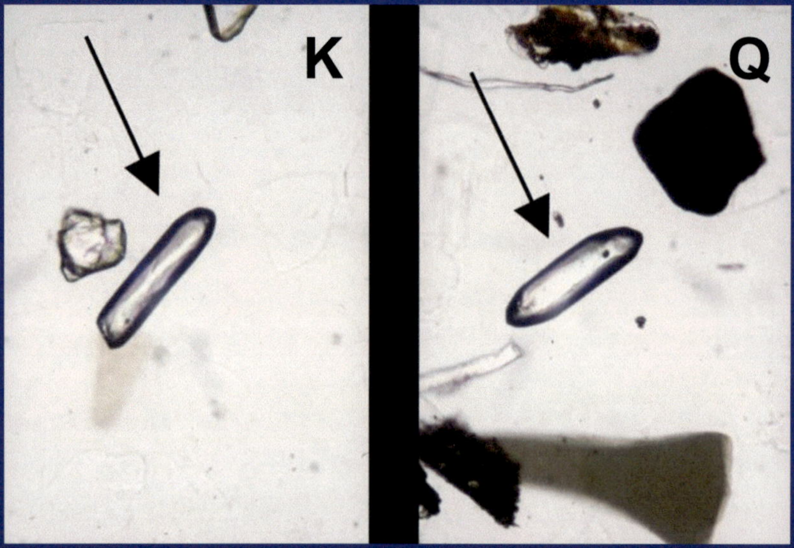

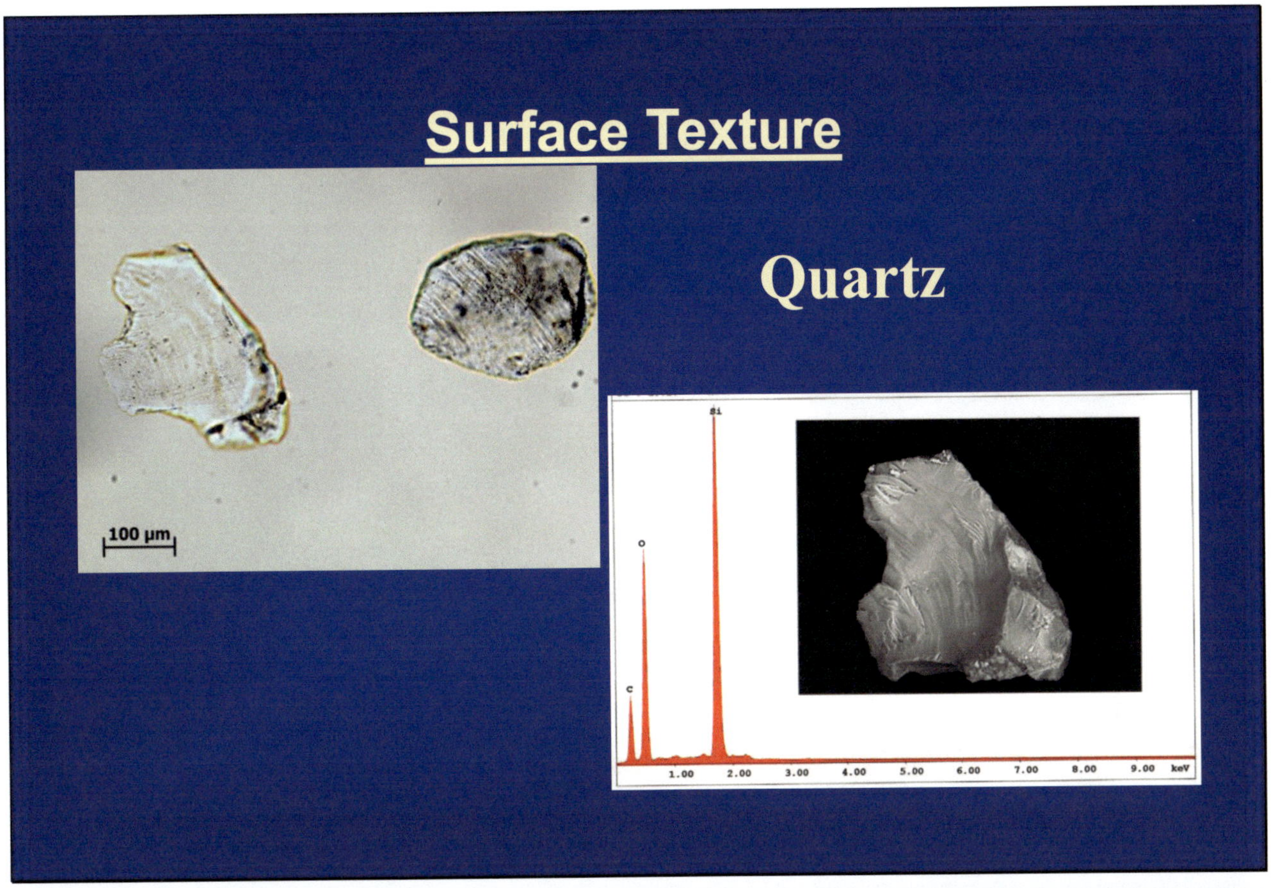
Surface Texture
Quartz
100 µm
Si
O
C
1.00
2.00
3.00
4.00
5.00
6.00
7.00
8.00
9.00
keV

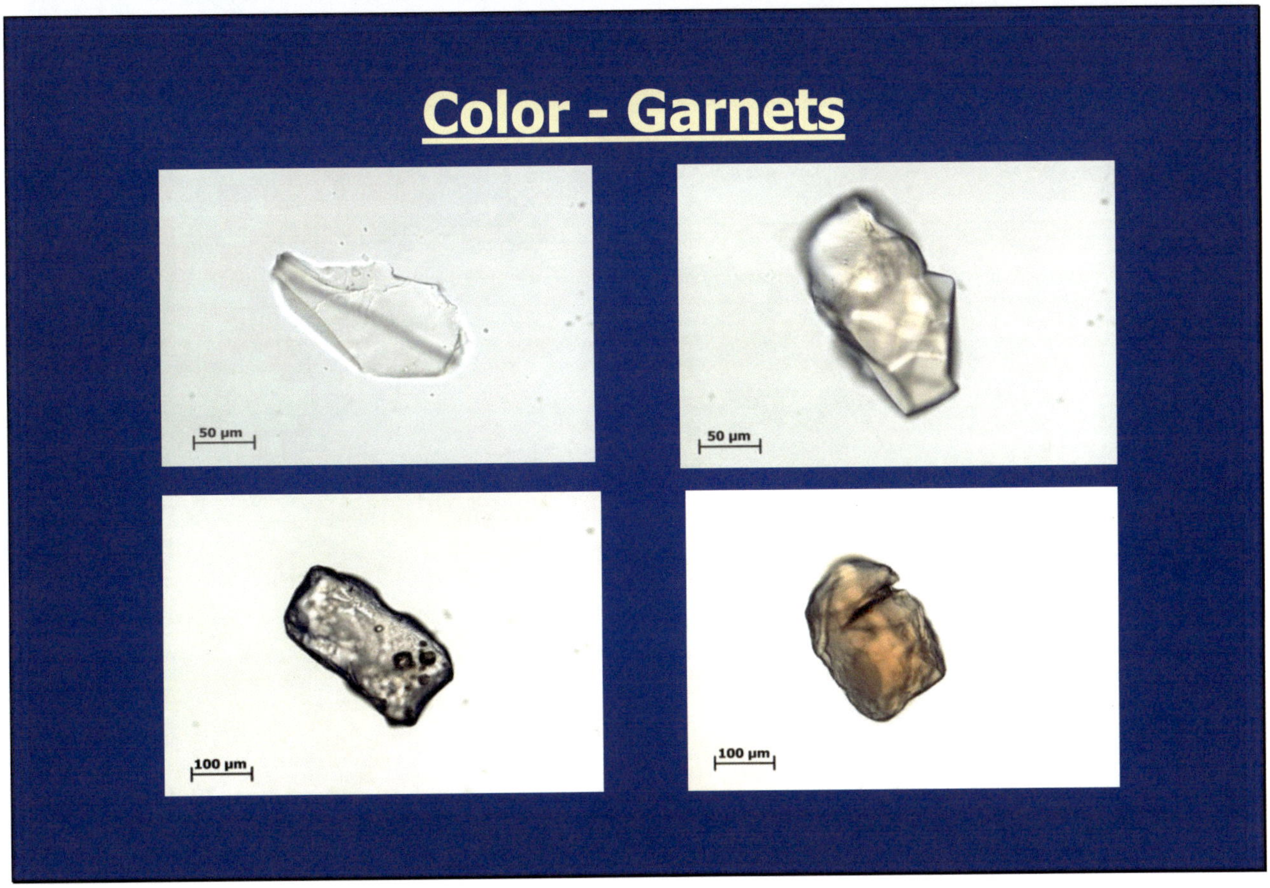
Color - Garnets
50 µm
50 µm
100 µm
100 µm

Color - Pleochroism

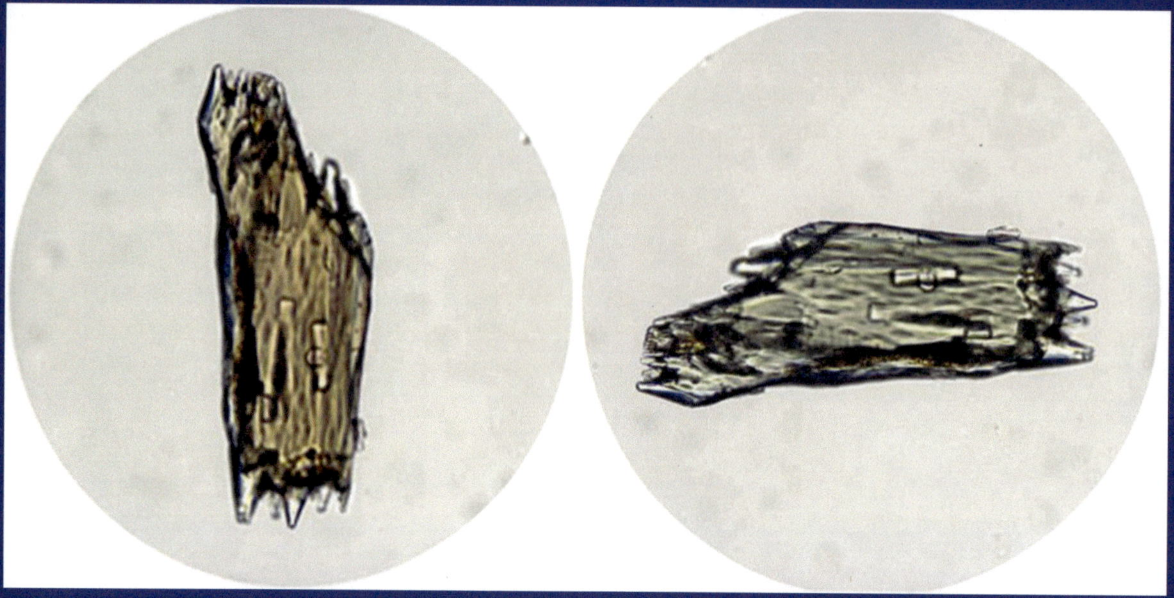

Color - Pleochroism

Hornblende

Relief

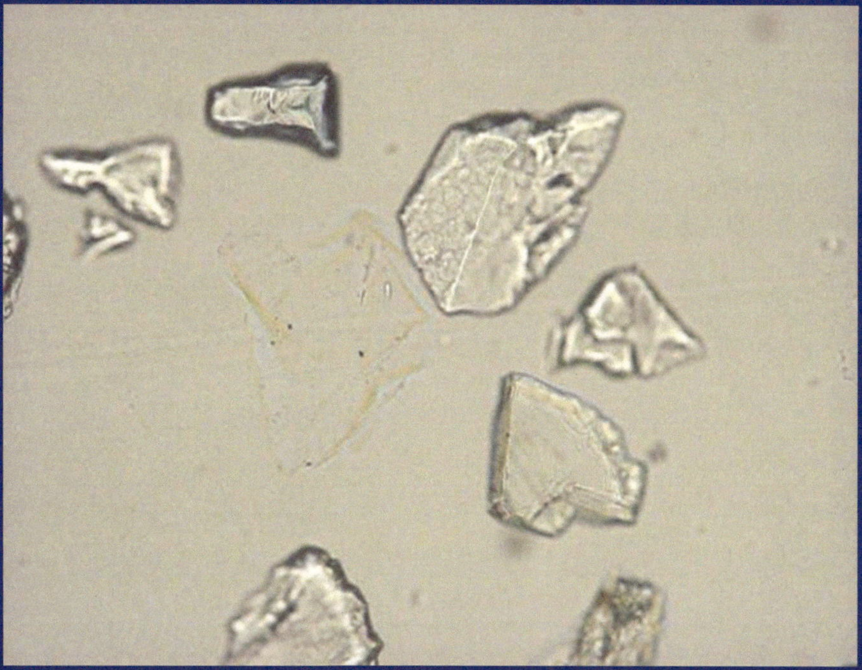

Refractive Index (Indices) - Crystal Systems

$$\text{Refractive Index (n)} = \frac{\text{Speed of Light in Vacuum}}{\text{Speed of Light in Transparent Material}}$$

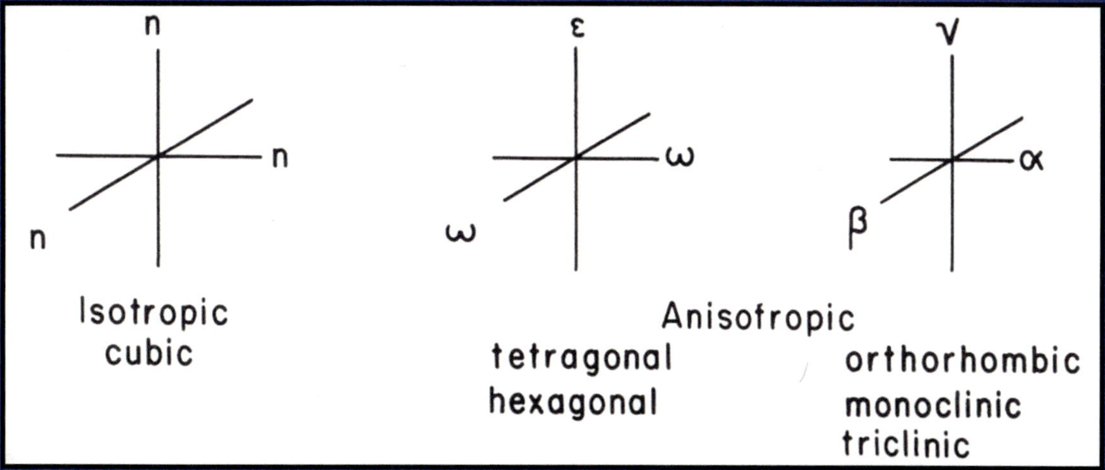

Refractive Index (Indices)

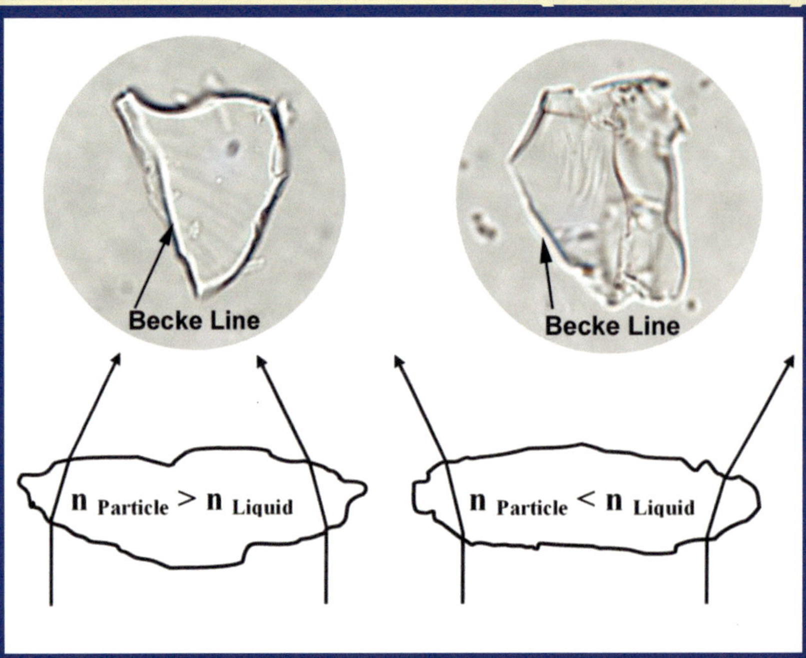

Retardation - Birefringence

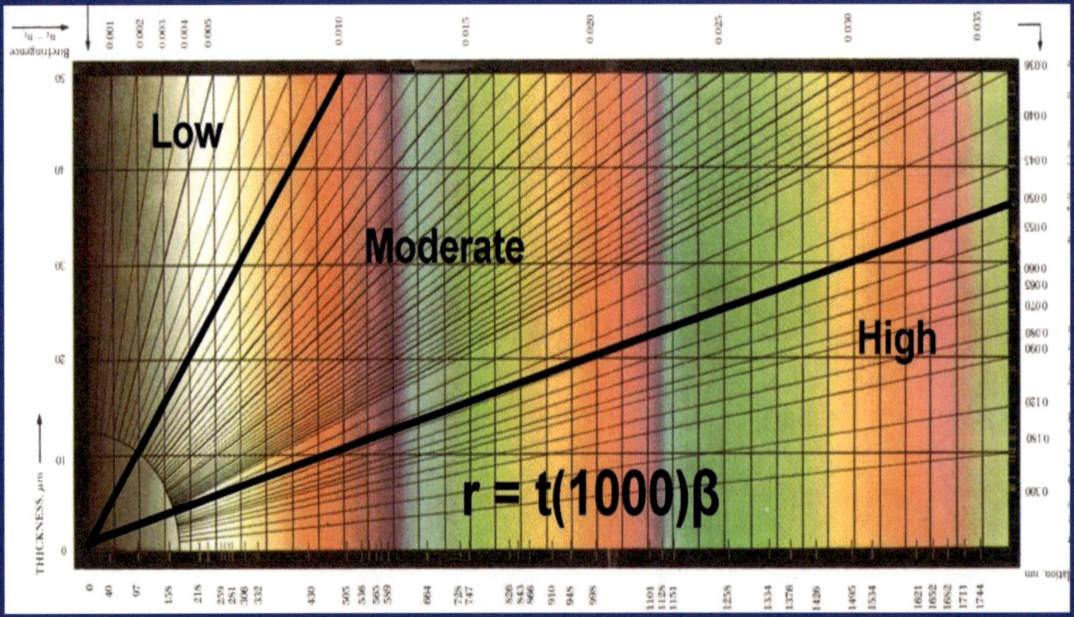

Interference Figure

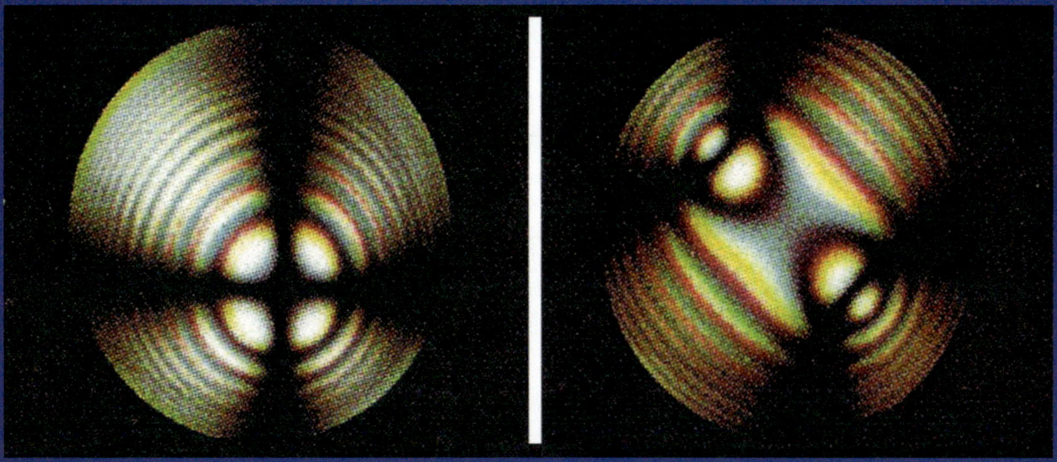

Interference Figure - Augite

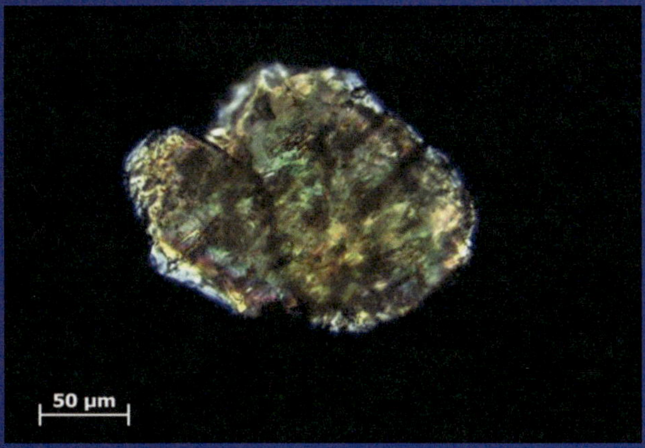

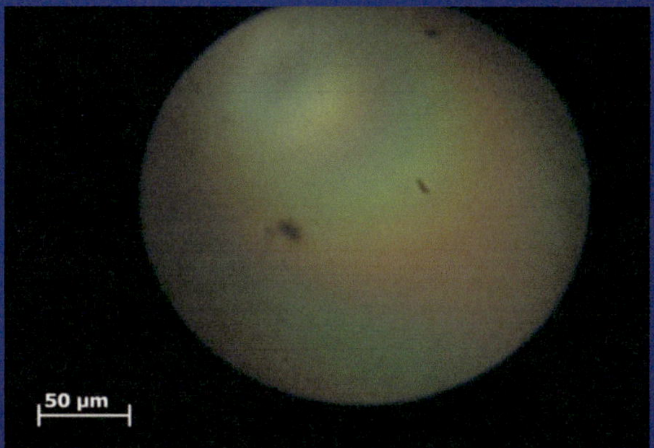

Extinction

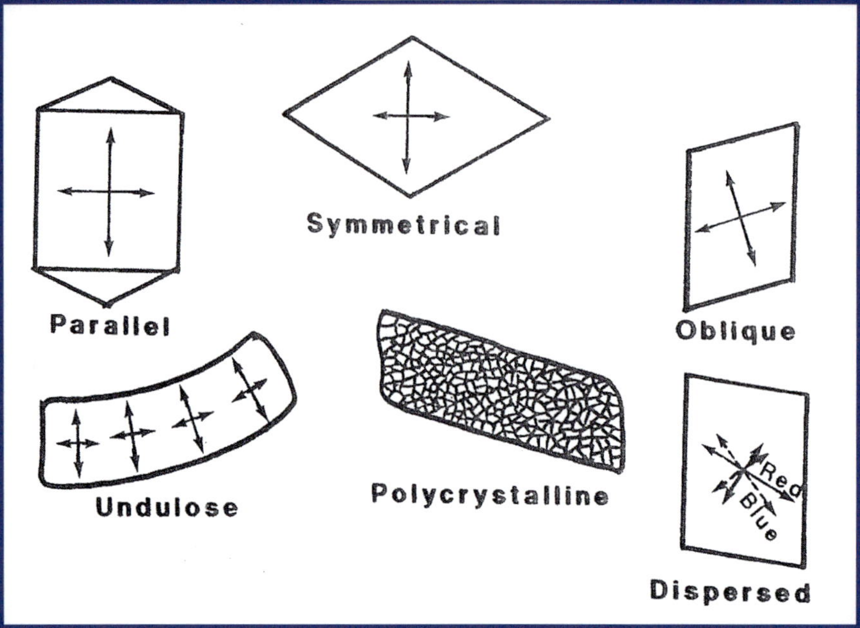

Twinning

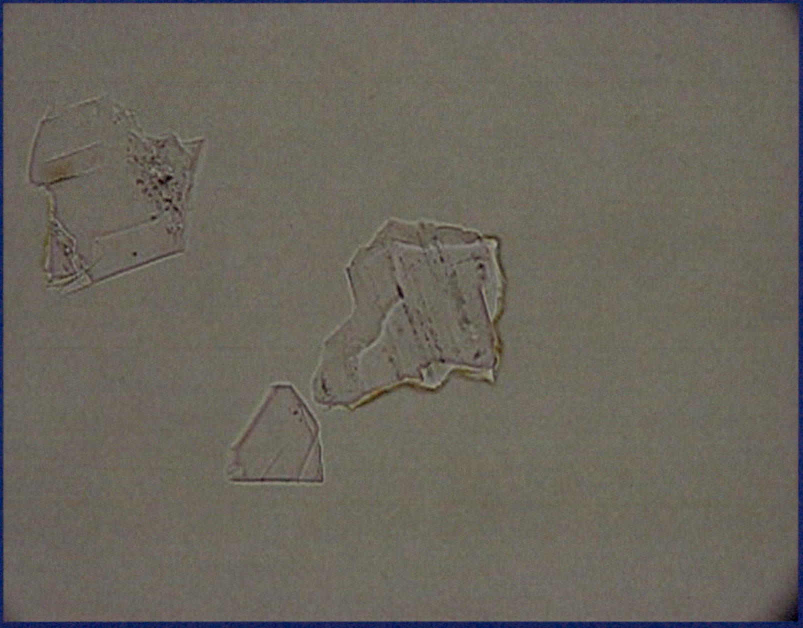

Inclusions

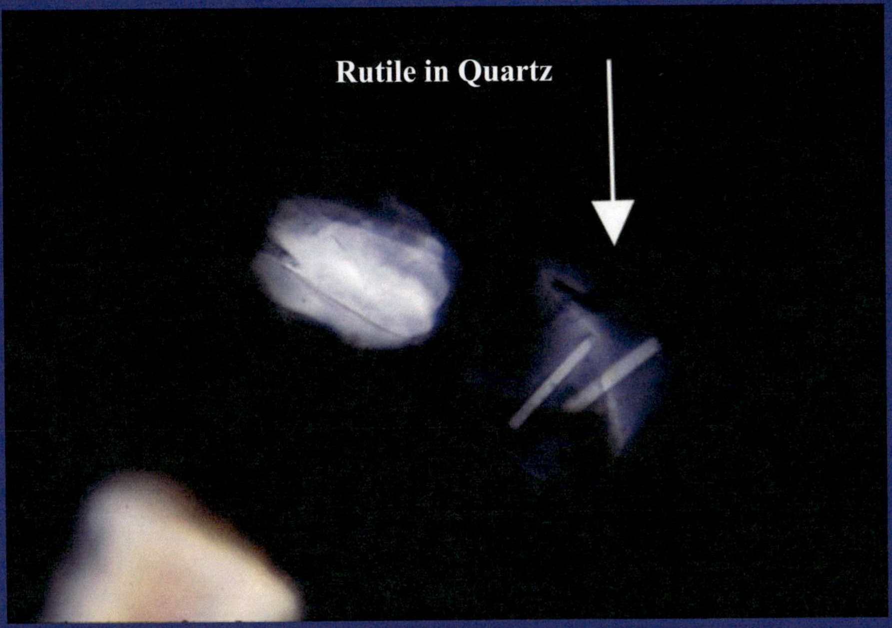

Inclusions - Garnet with Quartz

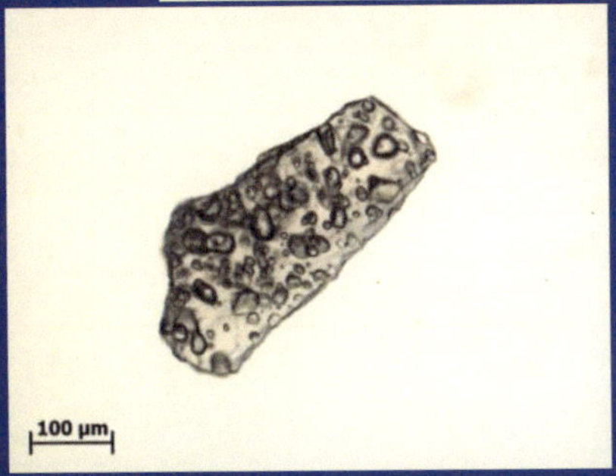

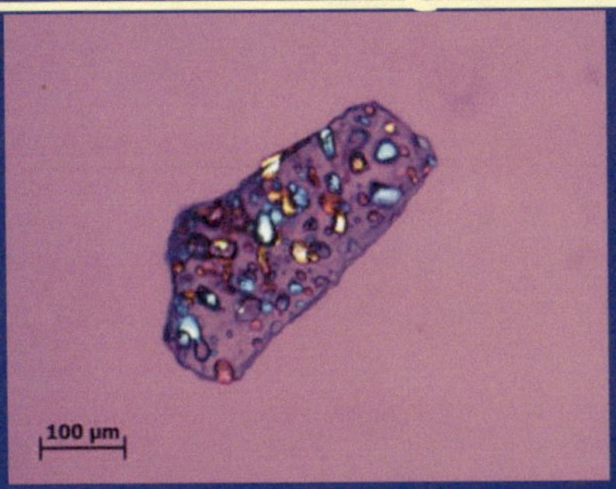

PLM ID Table

PLM Mineral Grain ID Reference Table

Mineral	Shape	Color/ Pleochroism	Contrast n vs. medium	Birefrin-gence	Extinction	Other (inclusions, surface features, interfere figure, optic sign, etc.) & additional (*) comments.
Light Fraction Mounting Medium: 1.545 HD						
Quartz	Angular Grains*	Colorless	Low Moderate n's - 1.544 & 1.553	Low	Complete	May have rutile needle inclusions. Conchoidal fractures common. Uniaxial (+). *Grains can be "rounded".
Alkali Feldspar	Tablets*	Colorless	Low n's < 1.540	Low	May show twinning	Biaxial *Usually show good cleavage
Plagioclase Feldspar	Tablets*	Colorless	Very low to low n's ~ 1.53 to 1.59	Low	May show twinning	Biaxial *Usually show good cleavage
Plant Opal	Varies	Colorless	n ~ 1.47	none	NA	
Calcite/ Dolomite	*	Colorless	Varies to v. high 1.48/1.50 to1.66/1.68	High to very high	Symmetrical	*Euhedral rhombs and irregular fractured grains. Uniaxial (-). Calcite normally in light fraction-dolomite normally in heavy fraction
Heavy Fraction Mounting Medium: 1.660						
Augite	Elongated worn*	Colorless to green with no pleochroism	Moderate n's - 1.66 to 1.76	Moderate-High	∠ > 30°	Clinopyroxene – Biaxial (monoclinic). May show optic axis figure. Show blue edges & may have inclusions * Also poorly irregular to rounded.
Biotite	Rounded Flakes	Brown Weak	Low β & γ = 1.57 to 1.68	Low	Complete	Biaxial – 2V + 0 to 25° (-) obtained on most plates. May look uniaxial where 2V is very small. β & γ in plane of flake
Chlorite	Rounded Flakes	Green	Varies n - 1.57 to 1.67	Low	Complete	Small 2V- usually (+) obtained on most plates. May look like biotite but (+) optic sign.
Epidote	Rounded	Faint yellow to green/yellow	Moderate n's > 1.7	Moderate*	Complete	Biaxial, 2V = 64-89° (-). *May show anomalous interference colors.
Garnet	Rounded grains*	Colorless, pink, salmon & orange	Moderate to high n's – 1.72 to 189	-	-	Green is usually spinal and easy confused as being a garnet. *Also, may see angular irregular grains.
Glaucophane	Elongated grains	Distinctive blue and violet colors	Low n's ≤ 1.67	Low/Moderate	Not ‖ < 15°	Amphibole Mineral
Hornblende	Elongated *	Olive to blue-green	Low n's sl. < & sl. > 1.66	Low	∠ < 25°	Amphibole mineral *Also angular to rounded.
Hypersthene	Elongated grains*	Pink/red to green	Moderate Show blue edges	Low	Parallel	Orthopyroxene - Biaxial (orthorhombic) * Ends may be pointed. Also stubby cleavage fragments.
Kyanite	Elongated /Bladed	Colorless	Moderate n's 1.71 to 1.73	Low/Moderate	Inclined	Triclinic. Grains my show conspicuous cross fractures.
Muscovite	Flakes	Colorless	Low β & γ = 1.55 to 1.62	Low	Complete	Biaxial – 2V = 30-47° (-) obtained on most plates. β & γ in plane of flake.
Olivine	Irregular grains	Colorless	Low n's - 1.6 to 1.9	Low to Moderate	Complete	Orthorhombic. Show brilliant interference colors. Rare as soil mineral but found in sand
Rutile	Irregular elongate*	Red to yellow color Faint to Strong	Very high n's – 2.6 to 2.9	High	Elongated Grains - ‖	Uniaxial (+) *Also rounded grains.
Sillimanite	Elongated grains	Colorless to rose Weak	Very low to low n's 1.65 to 1.68	Low	Elongated Grains - ‖	Orthorhombic
Sphene (Titanite)	Irregular grains	Yellow to brown Weak	Very high n's 1.9-2.0	Very High	Incomplete Due to Dispersion	Biaxial. 2V ~ 27° (+)
Staurolite	*	Colorless to brn/blk. Pleochroic in shades of yellow.	Moderate n's – 1.74 to 1.76	Moderate	Complete	Orthorhombic. Usually has inclusions. *Irregular to somewhat platy and rounded.
Tourmaline	Elongated to round	Colorless to brn/blk. Strong	V. low to moderate n's – 1.62 to 1.69	Low	Elongated Grains - ‖	Some show blue pleochroism colors. (-) sign of elongation. Tetragonal -Uniaxial (-) on particles that show color in all orientations.
Zircon	Euhedral Elongate*	Colorless	High n's - 1.92 to 2.0	High	Elongated Grains - ‖	Tetragonal *Also, elongated rounded.

Iron Oxides by reflected light: **Magnetite** - black; **Hematite** - usually black or red; **Limonite** - yellow and is usually a coating on other minerals.

T. Hopen - 02/2012

Bob Jarrett

Sand-making in Georgia: From the Tennessee Border to the Coasts

Current sand deposits have the potential to become future "ancient" deposits, whether loose or solidified to rock.

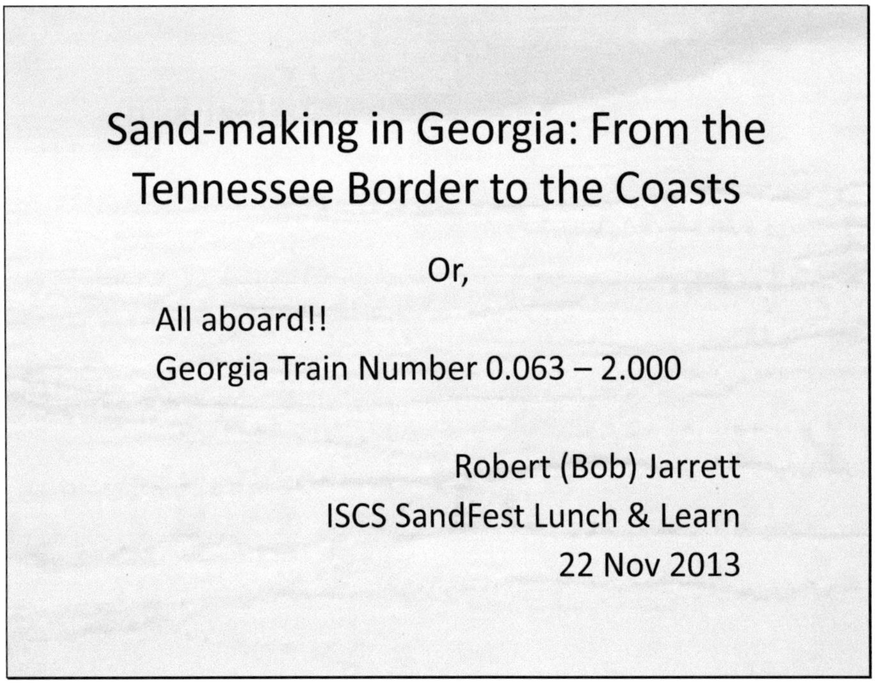

Take your shoes off, so we can walk in Georgia sands!

These are the sand facets we'll discuss. Starting, moving, ending.

Where Georgia sands lead us

- Sand origins - Georgia's five rock provinces' peculiarities
- Sand movement
  - Mechanisms
  - Directions
- Sand deposits -
  - Kinds
  - Locations

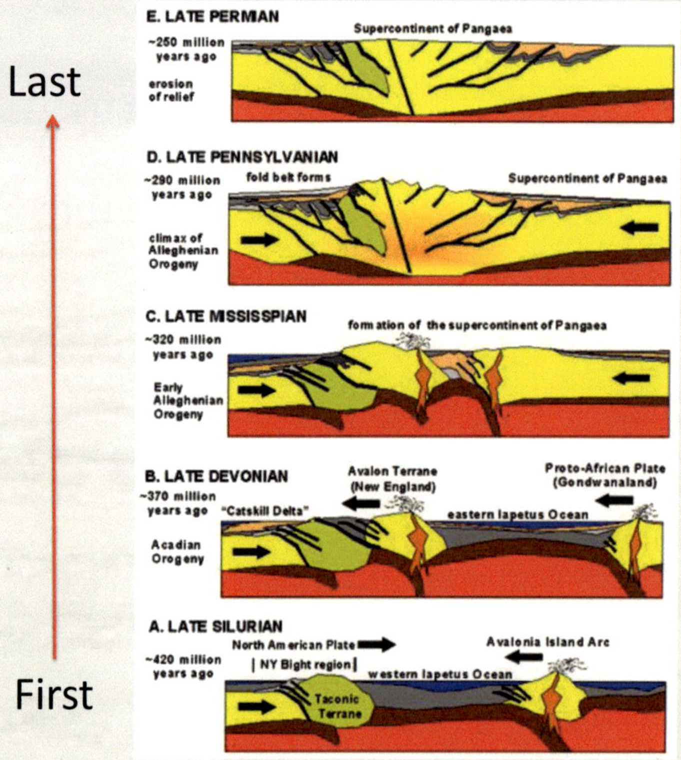

Source: Wikipedia (Jul 2012)

Take it from the bottom. As continents moved, they squashed each other. Mississippian inland sea deposits to Pennsylvanian to Permian to modern eroded.

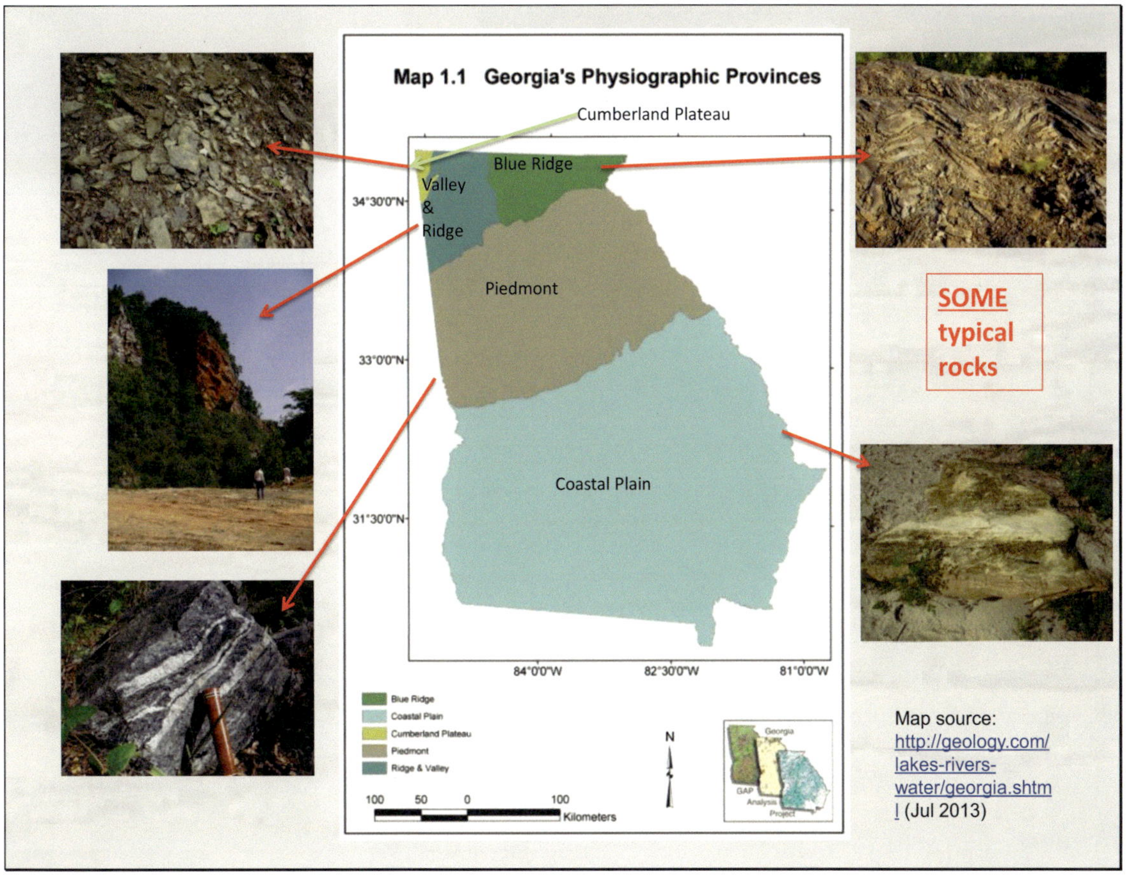

Georgia has five distinct geological "PROVINCES".

Cumberland Plateau: shallow ocean bottom and swamp deposits formed limestones, sandstones, shales, and coals.

Valley and Ridge: Deep ocean bottom sediments got squooshed up into blocks of sometimes vertical formations, in which some commercial mineral deposits formed.

Blue Ridge: Granites, limestones, and other rocks were pressed and heated as they rose to mountains as or higher than the Himalayas, so they "metamorphosed" into new, crystalline rock types.

Piedmont: Rocs of many types that were pressed, heated and changed to other crystalline rock types.

Coastal Plain: The "basement" of ocean bottom rock formed and modified by the collision of North America and Africa is overlain by as much as 3.5 miles of eroded sands and muds up to 80 million years old.

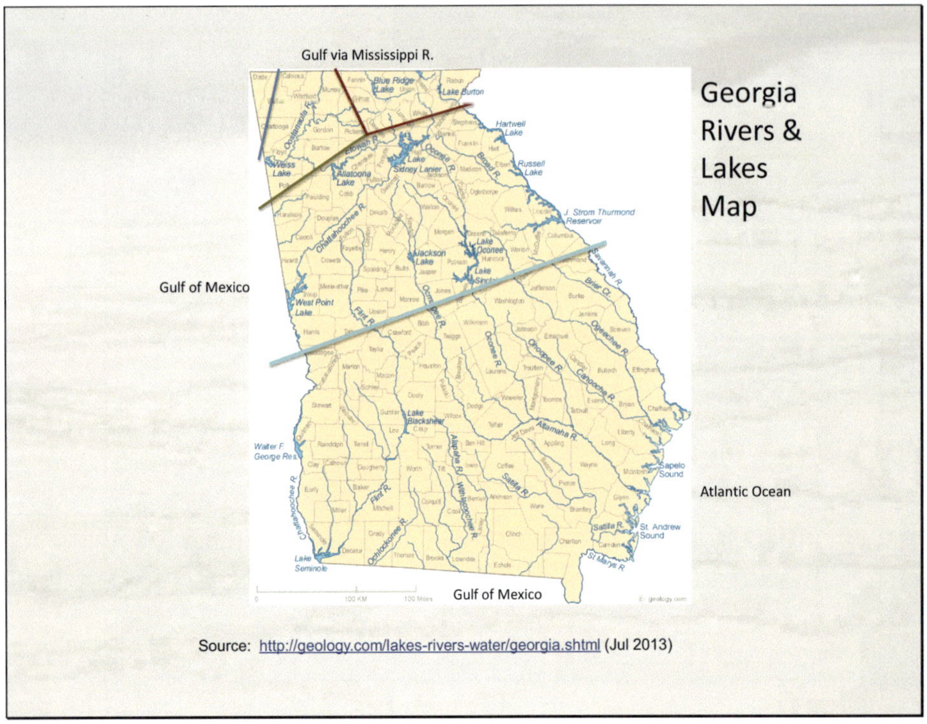

This is a quick look at the distribution of Georgia's rivers and major lakes.

The Blue Ridge Province stretching from Georgia up into Virginia exhibits steep, sharp ridges of hard rock. Valleys tend to follow fault lines between layers of relatively hard rock types, as the weakest, most easily eroded materials, where rock had been broken more by Earth's tectonic forces.

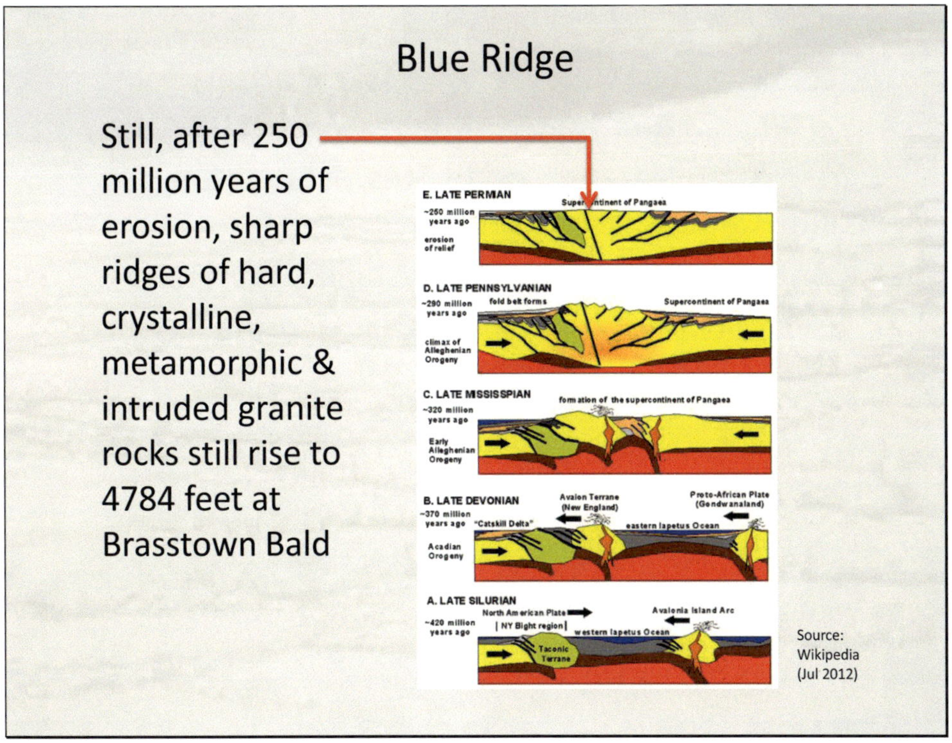

Still 250 million years after the end of the "smooth" Permian surface, we see sharp ridges and mountains. Up to seven mile-high mountain peaks are believed to have turned into sand, clay, and soil and washed away in those 250 million years. Brasstown Bald is still nine-tenths of a mile high above sea level!

Hard, crystalline rocks erode slowly and generate toll waterfalls. Water flowing down steep sloped ridges possesses great power to bash boulders and clean away resulting sand.

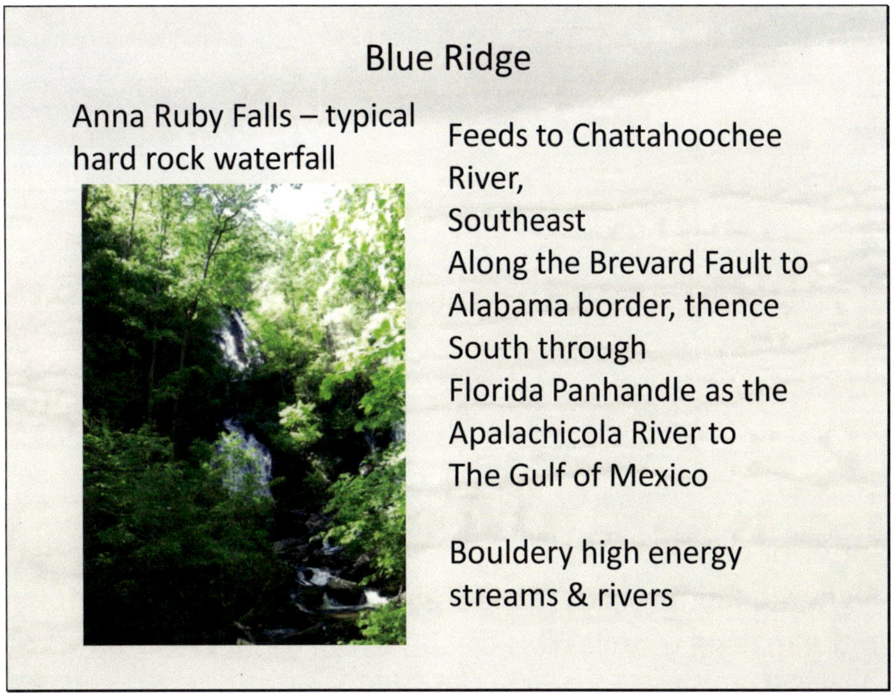

Blue Ridge Rocks

- Mostly metamorphic amphiboles, gneisses, marbles, pyroxenes, quartzites, schists, & some unmetamorphosed granites
- Contain a wide variety of crystalline minerals
  - ilmenite, magnetite, quartz, rutile, & zircon that make it to the sea – too physically & chemically resistant to be destroyed (More on these later)
  - Calcium carbonate, dolomite (chemically metamorphic), feldspars, garnet, hornblende, iron oxides, micas – & others too non-resistant to avoid solution or breaking to silt & clay sizes
- Gold – too dense to go very far; soft & erodes

Blue Ridge Rocks

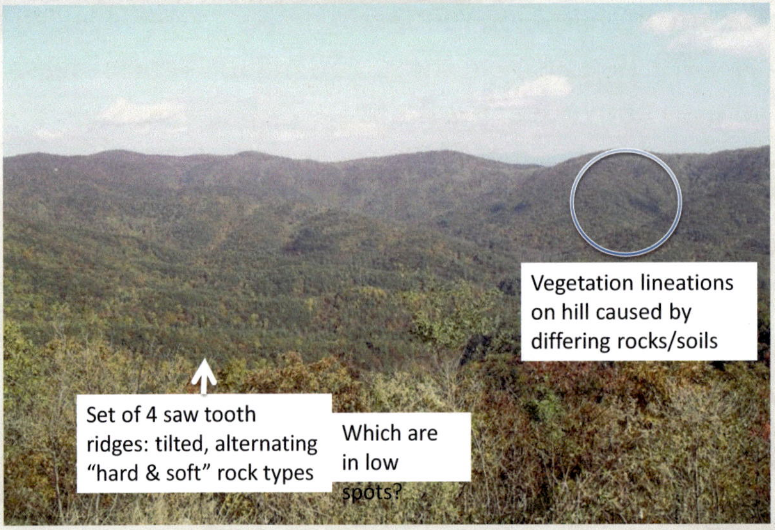

Softer or broken rocks are in the valleys which deepen most easily. Sometimes, visible vegetation differences result from the specific combinations of minerals that rocks release in the making of soils. This can be a tool for first estimates of the kinds of rock in those places, but that is a story for another time.

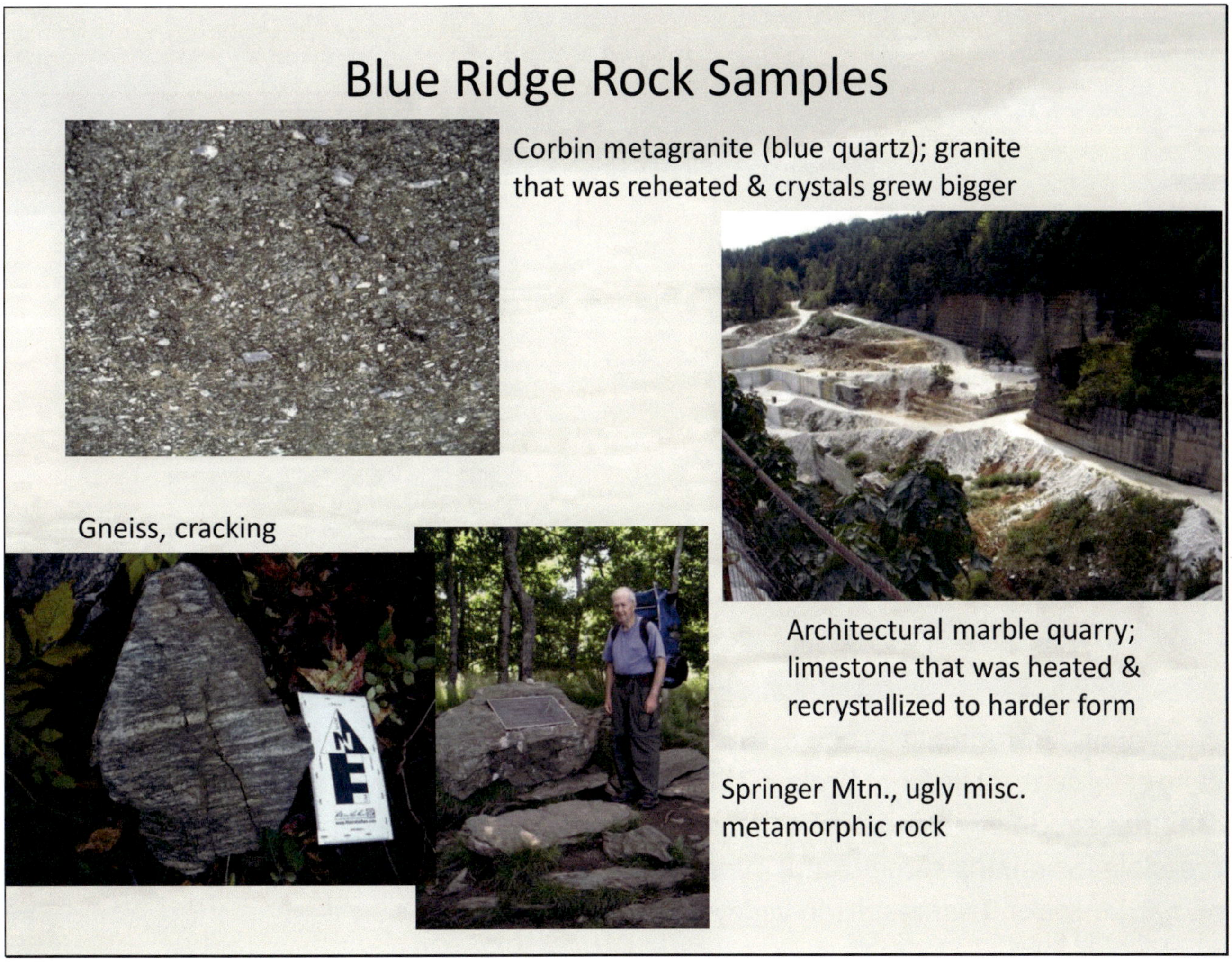

The Blue Ridge contains many kinds of metamorphic rocks:

Metagranite made by changing granite by high heat and pressure

Gneiss made from almost any kind of rocks and sediments by high heat, pressure, and (sometimes accompanied by) flow, and

Marble resulting from application of heat and pressure to soft limestone.
Heated water flowing through the rocks also dissolves and deposits many chemical elements and minerals during the overall process.

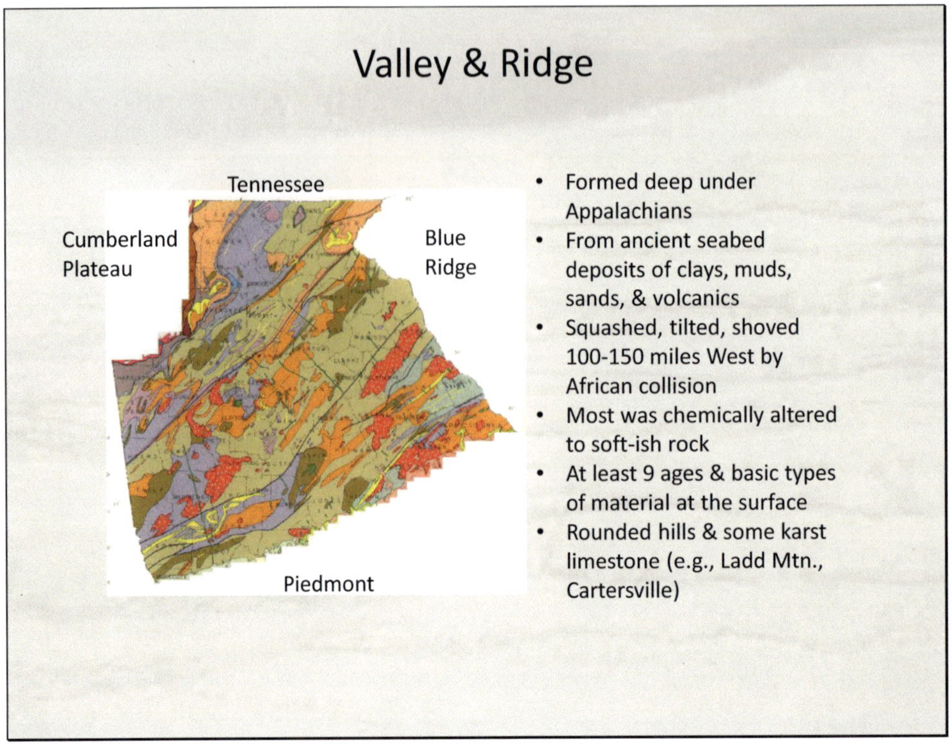

Colors signify differing ages and compositions of rocks at the surface. Large blocks of ocean sediment rocks were broken, tilted, and pushed partly over each other, like a row of dominoes set on edge and pushed to overlap each other. The edges are now visible and accessible for mining some useful minerals. This also exposes alternating stripes of softer and harder rocks. That results in <u>valleys</u> and <u>ridges</u>.

N 34° 10.536’, W 84° 45.233’ Highway I-75, Georgia Exit 288

From left to right or vice-versa, almost vertical layers show how much twisting and shoving occurred when the Allegheny Mountains were created. The original ocean bottom sediments were essentially flat to begin with!

N 34° 10.536’, W 84° 45.233’ Highway I-75, Georgia Exit 288

Valley & Ridge Minerals

- Construction grade limestone, barite, talc, iron & manganese oxides (for pigments)
- Noted chiefly for its production of brown iron ore, manganese oxide ores, barite, sienna, & magnesian limestone (dolomitic), brick & tile clays, slate, gold, bauxite, & graphitic schist.
- Minor deposits of serpentine, sericite schist, copper ore, & red iron
- Brown iron ore deposits are associated specular hematite

These tend to become particles of sand or silt and clay sizes.

Valley & Ridge

- Long gone dolomite caverns with barite covered in iron oxide ochres
- All chemically & physically soft

Valley & Ridge

The take away message:

- Most are relatively soft
- Or breakable
- Do not resist
- Might make sand immediately
- But rapidly weather to silt, clay, & dissolved chemicals
- Therefore, low rounded hills & ridges

Ladd Mountain, just west of Cartersville, Georgia is an exposed ridge of chemically soft limestone that used to be deeply buried with rivers running through it. Many fine examples of cave flowstone, stalactites, and stalagmites and late Pleistocene to Holocene Age animal fossils have been found there. It is no longer open to public access.

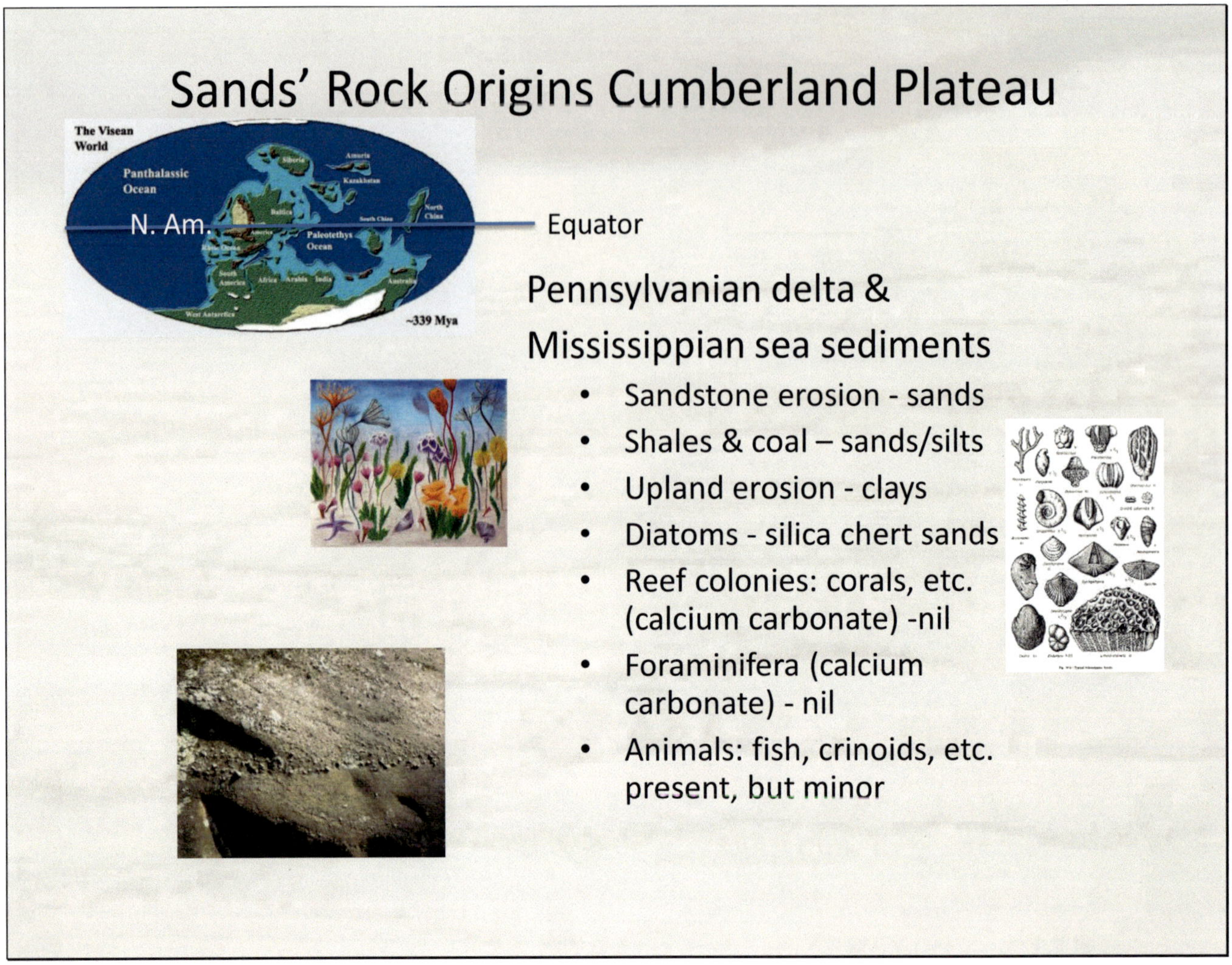

Plate tectonics is a mover and a shaker! In the Carboniferous Period (Mississippian and Pennsylvanian) North America was in the tropics, with huge river deltas and swamps along the shores of a great inland sea. Life was abundant and erosion carried much sand and other sediments to the seashores and nearshore ocean bottom.

The Cumberland Plateau contains horizontal layers of those materials, such as limestones, shales, coal deposits. There are also remnants of beach sands in the shapes they had then: layers, ripples, and solid beds. All layers are softer than Blue Ridge crystalline rocks, but some are harder than others. This results in some Sharp cliffs, deep canyons, and high waterfalls – the harder layers protect softer ones for a while.

These are examples of very thin, alternating coals and clays and carboniferous shale.

Piedmont

Medium hills & plains with gneiss bedrock outcrops common from North of Atlanta to the "Fall Line"

Piedmont

Slower rivers, with & without gneiss bedrock outcrops; plus sandy & rocky floodplains

Placid, sometimes, but boulder breaking ability, when raging

Piedmont Rock Types

Hard gneisses, pyroxenes, & amphibolites

Schist (this is garnetiferous)

Anomalous sandstone lifted in the Pine Mountain "Massif"

Schists and sandstones are easily susceptible to physical and other forces converting them to sand. Schists also chemically decay in a way that dissolves much of the mica content, as well as loosening the easily destroyed crystals to be reduced beyond sand-size.

Piedmont Minerals

Primarily

- Quartz
- Feldspars
- Micas
- Clays

Plus, everything from upstream not dissolved

Not a mineral, per se, but: Elberton architectural granite is quarried

These two rivers merge and eventually flow to the Gulf. The far bank is a silica sandstone with calcium carbonate cement holding it together. The acid swamp water corrodes the rock.

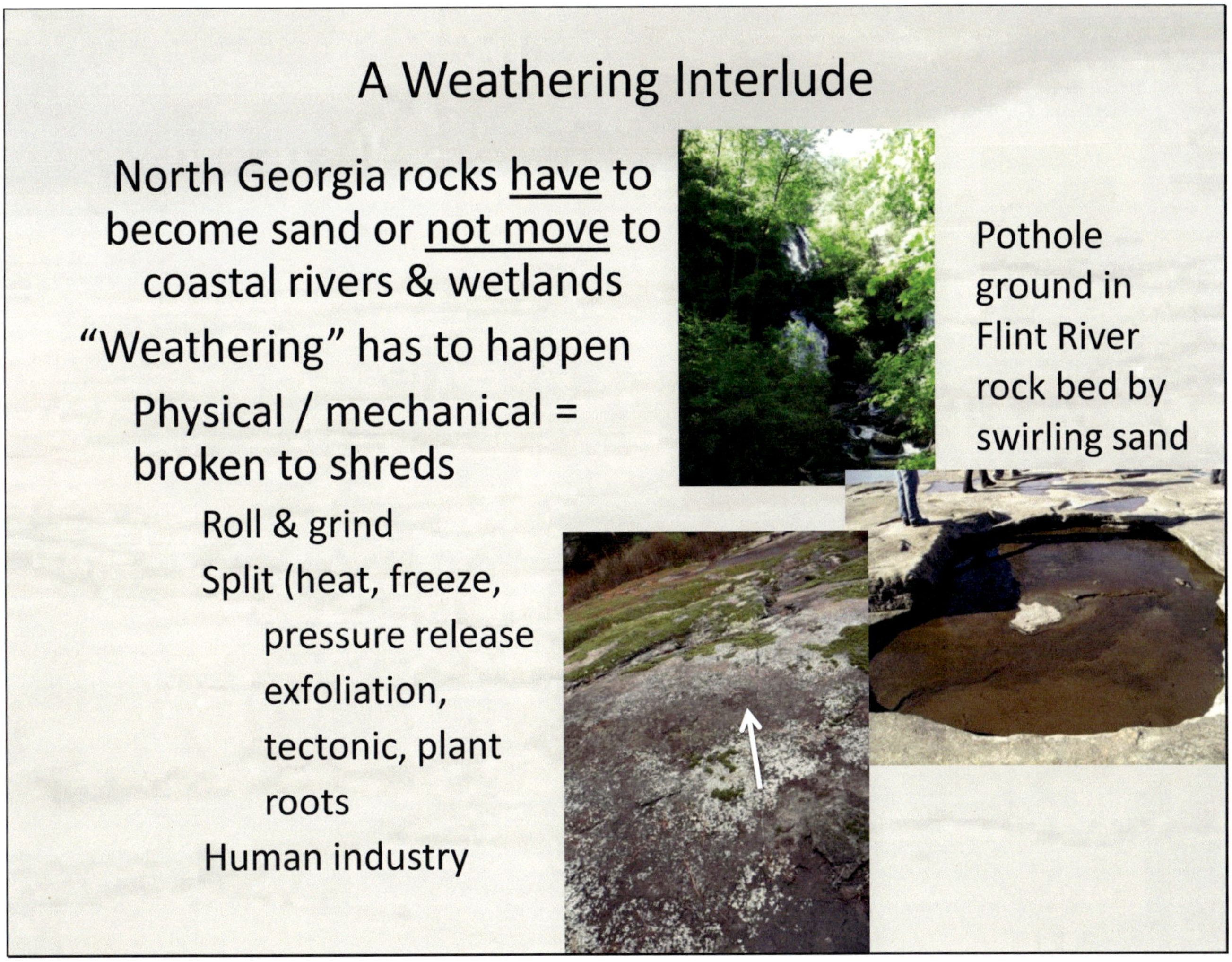

"Pressure release exfoliation" is a phenomenon of splitting and layers popping off of large rock masses owing to the release of stored pressure energy, when huge amounts of weight are removed by erosion and/or dissolution.

Arabia, Panola, and Stone Mountains Georgia and Yosemite's El Capitan are well known examples of where this happens. Big slabs of rock snap free like layers from an onion.

What is happening here?

What kind of weathering?

This is biological mechanical weathering, where the tree roots first entered a small crevice to obtain nutrients and water.

Then it grew and forces the rock to split, while more roots grow and soil forms in the large crack.

Not quite down to 2 mm sand yet, but the wheels are in motion.

The lower left picture shows three stages of chemical destruction of a particular kind of rock resulting in sand. The others demonstrate how rock structure may long last, as decay progresses.

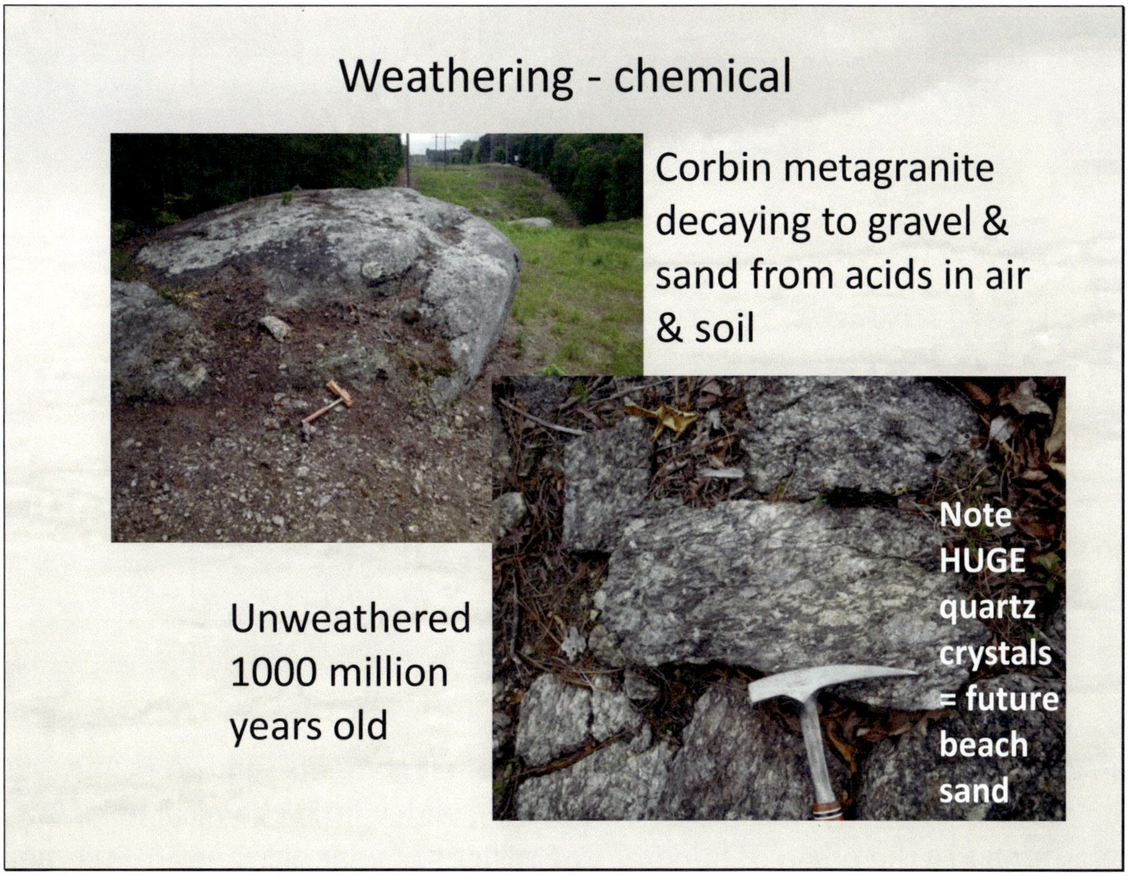

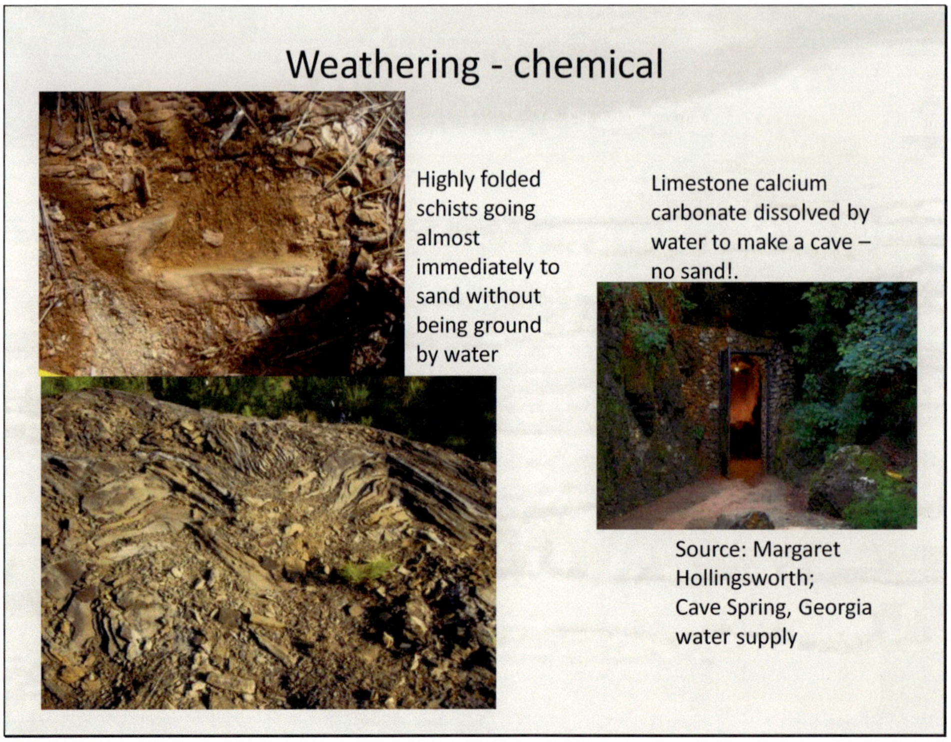

The rather coarse almost pure silica product is called "grüs". It contains sand-sizes and will eventually all be reduced in size by mechanical weathering with a little chemical help. But such material was probably the origin of hundreds of millions of year old sands from times when the only life on Earth was blue-green algae.

Small amounts of biological acids produced to leach nutrients from rocks make holes partially filled with loose quartz granules of various sizes. And, the acid attack can completely "rot" hard crystalline rocks, as well as softer ones. On the left, acid swamp water corrodes the rock along the Suwannee River of South Georgia. It leaves sand for this tiny beach to eventually be flushed to the Gulf.

Back to the Coastal Plain

Remember this map?

Coastal plain is the bulk of Georgia's area.

It is the sum of about 80 million years of weathering sediment accumulation

Sits on ocean-bottom-spreading basalt & diabase

In places, 3.5 miles deep

<u>Today's sand flows across</u>

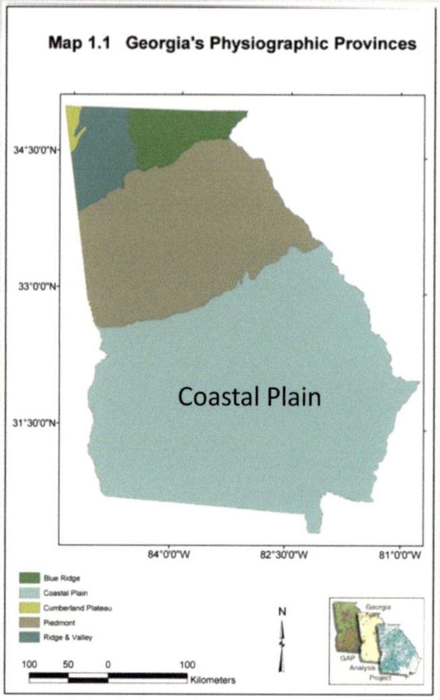

The coastal plain is a combined transport system, storehouse, and producer (form ancient sedimentary rocks) of sand for the Gulf and Atlantic Coasts.

Georgia Rivers & Lakes Discharge Direction Map

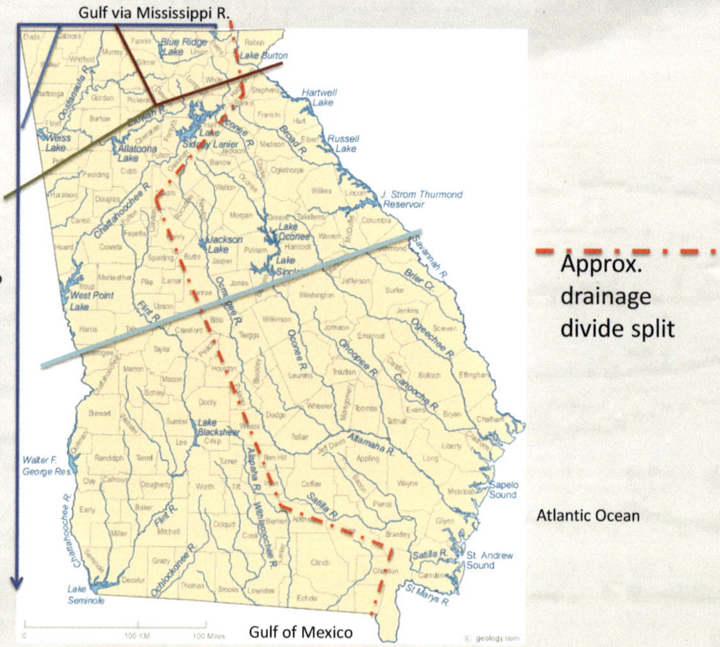

Source: http://geology.com/lakes-rivers-water/georgia.shtml (Jul 2013)

The dashed red line approximately shows the "continental divide" determining where a given drop of surface water or a grain of sand will go.

Coastal Plain

The sand comprises:

- Sediments from the ancient Appalachians
- New sediments from the North being made all the way along the rivers
- Sediments released by erosion of Coastal Plain formations
- Sand blown from ancient rivers on the West, as at Ocmulgee State Park

Exposed Ocmulgee sand dune

Coastal Plain

Warm Earth periods meant high sea levels with beaches forming far inland up to 50 meters higher than at present

Trail Ridge at Okeefenokee Swamp is such a beach, rich in ilmenite, zircon, rutile, epidote, sillimanite, staurolite, magnetite, tourmaline, & kyanite

Titanium is in ilmenite & rutile

The black streaks are black "heavy minerals", such as magnetite. Its visible presence can be used as a clue that more valuable heavy minerals might be present in a sand body. Rutile is titanium dioxide. It is an important industrial chemical for making titanium metal, sunscreens, white paint pigment, and whitening ingredients for many other products. The others have major beneficial uses, also.

In review, when all is said & done:

About all that gets to the beaches as sand are:

ilmenite, ← titanium-iron oxide
magnetite, ← iron oxide
Quartz, ← The overwhelming fraction, by far
rutile, & ← titanium dioxide
zircon ← zirconium silicate– too physically & chemically resistant to be destroyed through billions of years

Plus, whatever sea & freshwater fossil materials grew in the coastal plain during ages of sea level rise & fall & animals that grow now

These are the heavy minerals found in Georgia sands.

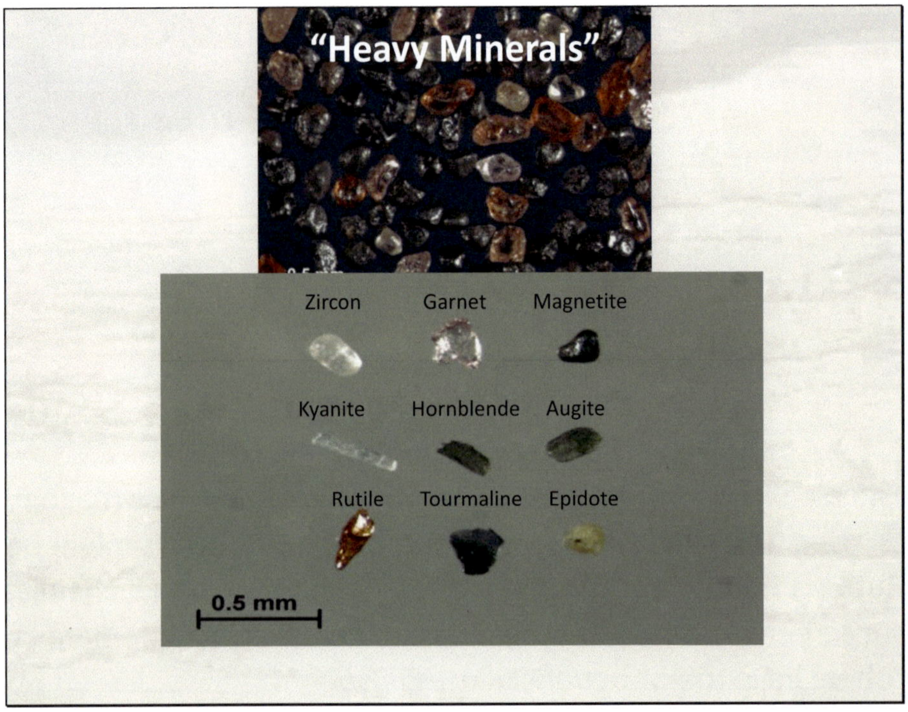

These sands were deposited during the Cretaceous period before the dinosaurs' demise. A variety of colors exist in this region, pretty much within arms' reach of each other. Many people use them for artistic purposes, like sand jars and paintings.

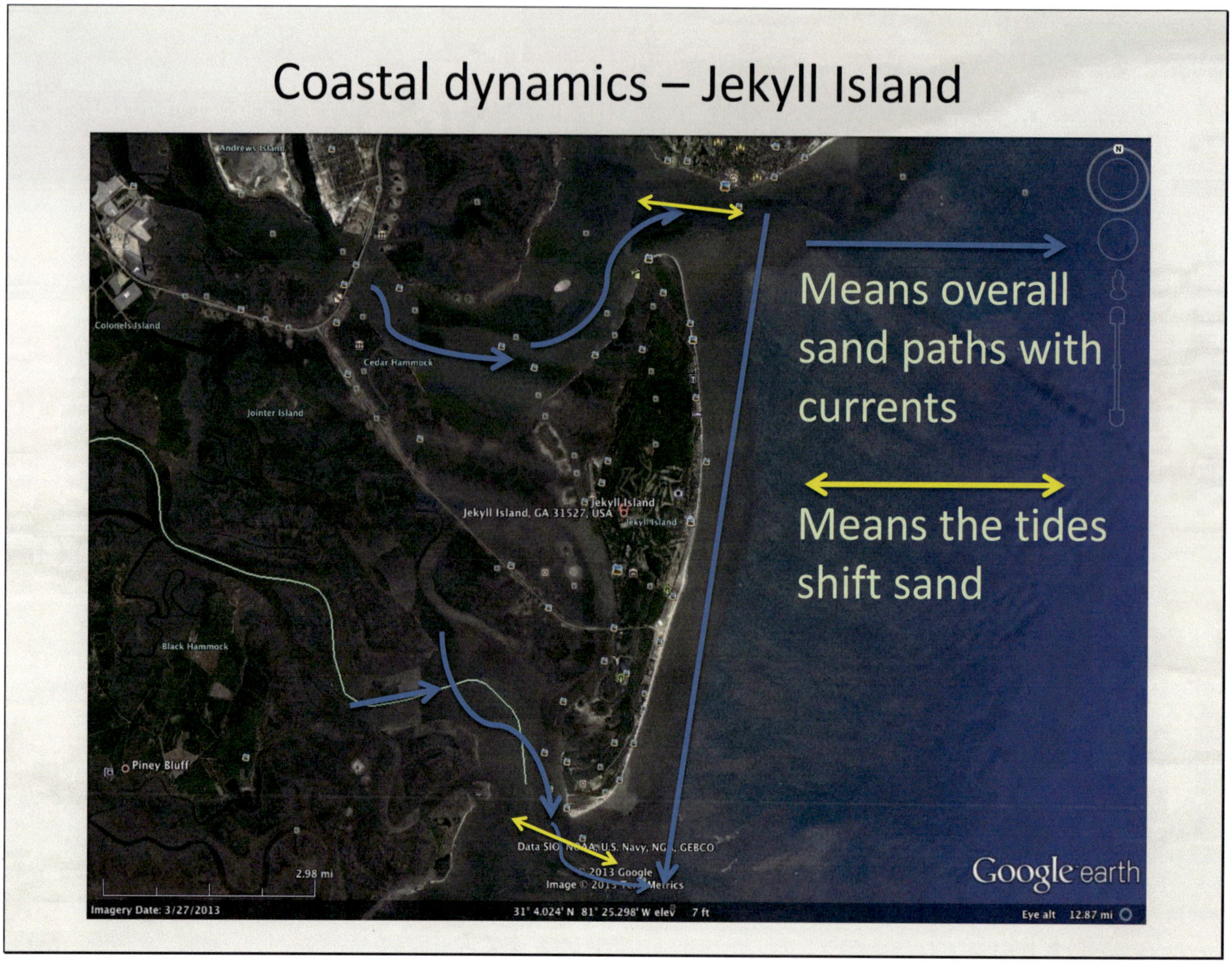

Sand approaches the seashore in rivers. It may temporarily lodge in sloughs and lagoons or be swept immediately to the surf.

Tides can return sand partway inland.

Over time, a given particle can expect to move along any given coast according to prevailing currents, while experiencing innumerable shifts from beach to sea bottom.

Some sand might be blown partway inland curing dune building processes.

If sea levels drop, dunes and sand islands may harden (lithify) into new sandstone, or they may be carried further seaward. If sea levels rise they may become water-covered and buried in the seabed under other sediments to possibly become new sandstone.

No rest for the weary!!

Unlike this picture, most Georgia sea beach sand is silica sand with some other hardy minerals, but shell building animals produce long-lasting shells and shell fragments that mix with the hard mineral sands.

So, Beaches are rarely pure anything.

Please, let's relax in the artificial sand bar.

What more should we talk about?

Ralf Hermann

Sand Collecting in Europe:
The History and Practice Today for the German Sandmuseum

1985 The german Horst Windisch foundet the SCI – Sand Collector International in South Africa (50% US-american Members)

1985 The first Sandjournal „The Sands of Time“ was published

1991 The first Sandcollector Meeting in the world took place in Pforzheim/Germany

1985 The german Horst Windisch foundet the SCI – Sand Collector International in South Africa (50% US-american Members)

1985 The first Sandjournal „The Sands of Time“ was published

1991 The first Sandcollector Meeting in the world took place in Pforzheim/Germany

1994 Joop Van Dyk organized the first International Sandmeeting in the Netherland /Wyk aan Zee.

1985 The german Horst Windisch foundet the SCI – Sand Collector International in South Africa (50% US-american Members)

1985 The first Sandjournal „The Sands of Time“ was published

1991 The first Sandcollector Meeting in the world took place in Pforzheim/Germany

1994 Joop Van Dyk organized the first International Sandmeeting in the Netherland /Wyk aan Zee.

1995 Some Austrian Collectors met for an one-off Sandmeeeting.

2004 The french Sandcollectors founded „Association Francaise d’Arenophobile and there Bulletin.

1985 The german Horst Windisch foundet the SCI – Sand Collector International in South Africa (50% US-american Members)

1985 The first Sandjournal „The Sands of Time“ was published

1991 The first Sandcollector Meeting in the world took place in Pforzheim/Germany

1994 Joop Van Dyk organized the first International Sandmeeting in the Netherland /Wyk aan Zee.

1995 Some Austrian Collectors met for an one-off Sandmeeeting.

2004 The french Sandcollectors founded „Association Francaise d’Arenophobile and there Bulletin.

2007 Van Dyk died. His Collection went to Lommel/Begium where some Sandmeeting was following….and died also in 2012

1997 The Association German Sand Museum founded

2006 The Internet Sand forum went online

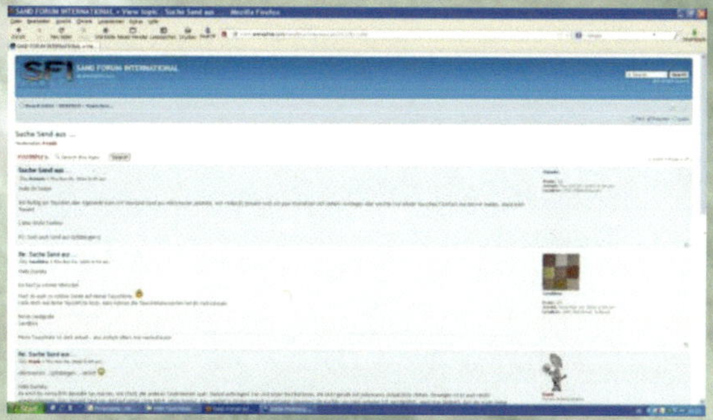

1985 The german Horst Windisch foundet the SCI – Sand Collector International in South Africa (50% US-american Members)

1985 The first Sandjournal „The Sands of Time“ was published

1991 The first Sandcollector Meeting in the world took place in Pforzheim/Germany

1994 Joop Van Dyk organized the first International Sandmeeting in the Netherland /Wyk aan Zee.

1995 Some Austrian Collectors met for an one-off Sandmeeeting.

2004 The french Sandcollectors founded „Association Francaise d'Arenophobile and there Bulletin.

2007 Van Dyk died. His Collection went to Lommel/Begium where some Sandmeeting was following….and died also in 2012

1997 The Association German Sand Museum founded

2006 The Internet Sand forum went online

2009 The exhibition Sandwelten open her doors.

SAND museum

1985 The german Horst Windisch foundet the SCI – Sand Collector International in South Africa (50% US-american Members)

1985 The first Sandjournal „The Sands of Time“ was published

1991 The first Sandcollector Meeting in the world took place in Pforzheim/Germany

1994 Joop Van Dyk organized the first International Sandmeeting in the Netherland /Wyk aan Zee.

1995 Some Austrian Collectors met for an one-off Sandmeeeting.

2004 The french Sandcollectors founded „Association Francaise d'Arenophobile and there Bulletin.

2007 Van Dyk died. His Collection went to Lommel/Begium where some Sandmeeting was following….and died also in 2012

1997 The Association German Sand Museum founded

2006 The Internet Sand forum went online

2009 The exhibition Sandwelten open her doors.

2009 The International Sandmeetings from the DSM started

SAND museum

© DSM – R. Hermann - Sandfest Cartersville 2013

The German Sand Museum and the practice of Sandcollecting in Europe today

2009 - 2013 The German Sandmuseum organize the International Sandmeeting in Europe yearly every second weekend in September.

Every year we try to find locations with „nice“ places where we could find interesting samples.

The locations in the surroundings of the meetings could show that you can find samples which are nicer and magnificant color as you normally knows german samples from the sea.

Colorsands Duingen / Lower Saxony

Mix from Clara Mine / Baden Württemberg

Garnet Topazolit St. Andreasberg / Lower Saxony

We try to inform about the technology of collecting samples

How to handling during journeys and what you need

Where you could find interesting material and what to do on the location.

We try to inform about the
technology of collecting samples

How to get concentrates during the journey

How to get Foraminiferas and Heavy Minerals from soaps on the Beaches

Knowing how = Better samples
Wherever you collecting

Europe are smaller then America.
But the multitude of small bays and beaches offer an enormous potential of different samples.

Saphir & Garnet / Bretignolles-Sur-Mer /France

Amber concentrate /Denmark

Ooide / Cleopatra Beach /Turkey

What's going on after the „collecting summer"

Normally the most collector have the most samples (for trading too) after the summer holidays. This ist the reason why you should ask for trades after July….

…or visit the european Sand Meetings.

„Dutch Trading"

The international standard of trading are welknown with 1:1.

The Dutch trading means that everybody may! place his samples on the table and everybody may take this samples which he/she believes need to be.

Before and after

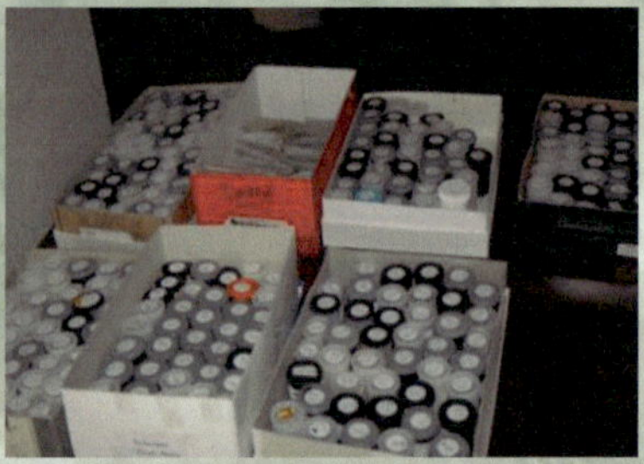

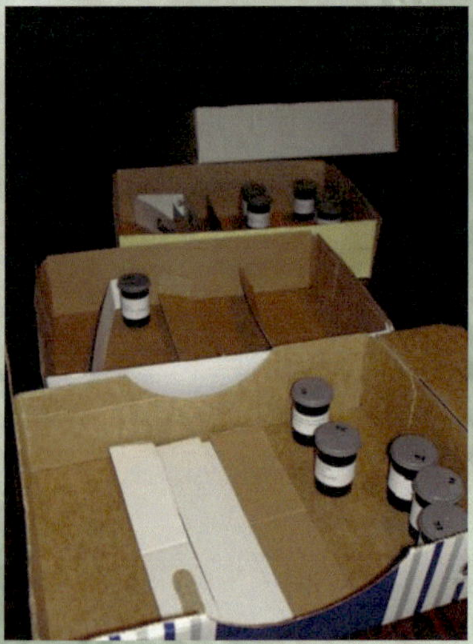

What the German Sandmuseum is doing in the next future in Europe?
- The International Sandmeetings in Europe
- The Sandjournal (English version of DSM Rundbrief)
-The IBI (International Beach Index)
-The Forum (German and English; the continuation of the old Sand Forum International)
- The standard list of sandlocations
SAND Journal
The World Provinces Index
SFI
SAND FORUM INTERNATIONAL
at arenophile.com
Skip to content
SAND museum
© DSM – R. Hermann - Sandfest Cartersville 2013

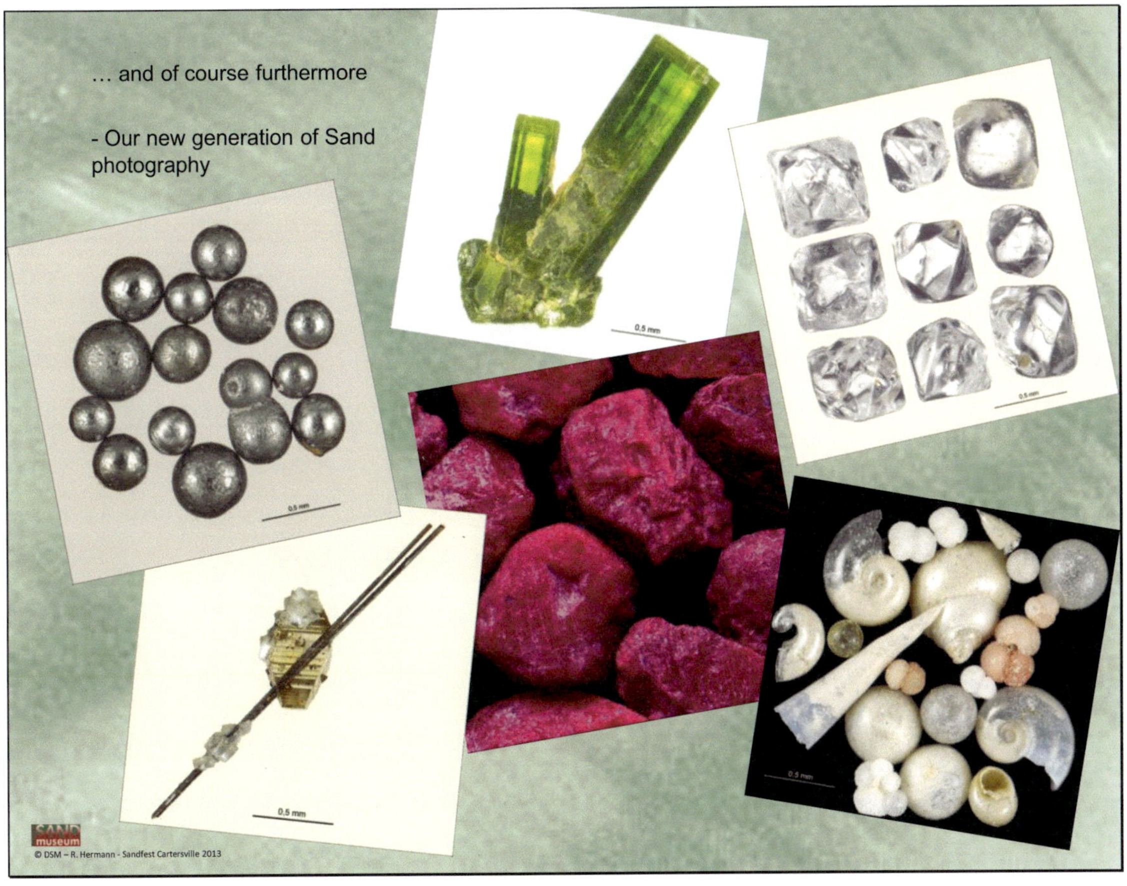
… and of course furthermore
- Our new generation of Sand photography
0.5 mm
SAND museum
© DSM – R. Hermann - Sandfest Cartersville 2013

… and visit friends…so…
…warm welcome during a visit to
Europe too; welcome to a visit to
the
deutsches SAND museum
Thank you for attention!
© DSM – R. Hermann - Sandfest Cartersville 2013

Anita D. Westlake

Sand Containers

Since our ancient relatives first crawled out of the murky depths, we've needed containers. Perhaps, that's why we left the water in the first place; in search of the ideal pocket or purse to keep our precious pebbles in.

Women, more so than men, seem to love the occasional box, bag, or bin. It's in our nature. Cave women needed containers to keep their sharpened bones sharp when not in use. Gatherers (of the Hunter-Gather fame) needed containers to carry food back home. Piling apples and grain on the top of the head or under the arms was unsightly, cumbersome and affected the taste in a bad way. The Egyptians needed containers to keep their perfumes, crushed beetles and berries safe from the arid winds of their native lands. Suburban housewives of the 1950's made Tupperware a household word. Men are still scratching their collective heads over that one.

And so, we jump ahead to modern homo-sapiens who need, want and love containers to keep the debris of life, in a word, *contained*.

But what of the 21st century collector? And more to the point, the 21st century *female* collector of things small and shiny. Here begins the tale of the magpie and the sand collector.

Magpies have been known throughout history as being avid collectors. A shiny bit of tin foil carefully arranged, along with a ribbon, button, or tiny rock do more than decorate the nest. It is the decorator's hope that a potential mate will see the finery and come in for a closer look. And isn't that what we, as collectors, are trying to do too? *"Please, come inside and see my etchings or at least my sand collection".*

Sand comes in a baffling array of colors. Even the sand from one side of a creek or riverbank can differ significantly from the opposite shore. In order to collect, clean and categorize our treasures, we must first contain them.

Materials can range from plastic, glass, wood, metal, ceramics, etc. When choosing the right material, consider the following factors:

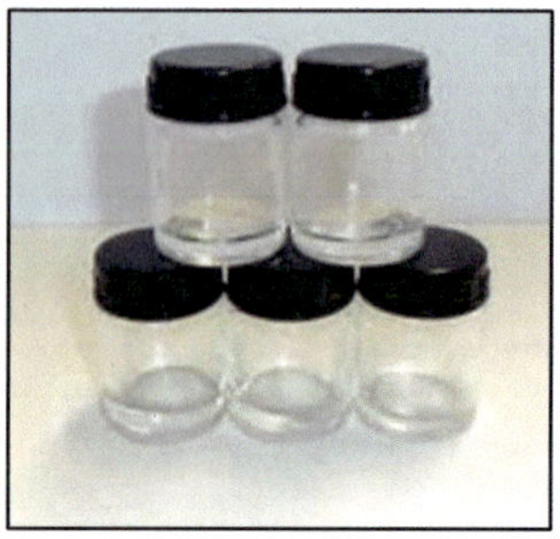

Glass, though durable, can be scratched by sand.

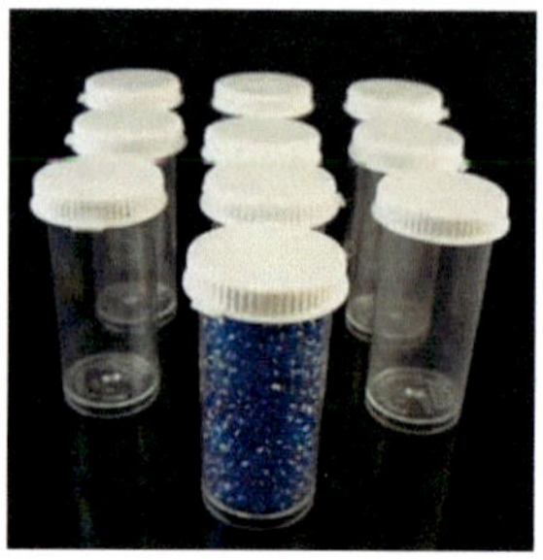

Plastic, though usually inexpensive, can be scratched or broken.

Transparency is a must. It definitely kills the mood if you have to struggle with opaque boxes and bags to show off your favorite samples.

Lids must fit securely to minimize spillage and contamination.

Choose a smooth surface where labels can readily be applied.

Size is important (contrary to what you may have heard). Will you be collecting small samples to view under a microscope? Large samples to share with friends? Both?

Decide the scope and purpose of your collection. Tailor your containers to fulfill that purpose. A collection that is intended mainly for display is often attractive if the same type containers are used. I use the plastic "gem jars" one can purchase at rock and mineral shows or the occasional hobby/lapidary shop. They fit neatly and snuggly into a foam base which in turn fits into a flat plastic holder. Fifty covered gem jars fit into one flat. All the subtle variations in color and size can be viewed at a glance, or taken out individually to be studied in detail. Labels can be made to fit on the back, thus keeping the top view unobstructed.

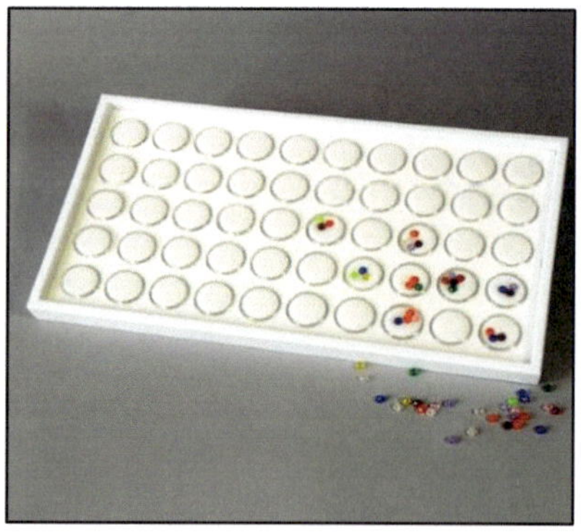

Other collectors use glass jars or tubes with cork stoppers. Keep in mind that cork is not durable enough to suffer multiple openings and closings.

Metal containers may erode over time, depending on the sand's makeup and moisture content.

Wooden containers are not a good choice for two reasons: obstructed view and out-gassing. Certain woods, and certainly the glues used to construct the containers can out-gas vapors that could prove to be detrimental to your treasured samples.

Clear, plastic containers with magnifier tops are fun to use but are very small and expensive.

Small, plastic bags should only be used temporarily. They will cloud over with handling and are not durable enough to withstand the movement of the sand inside.

Consistency in the size and type of container usually creates the best visual effect. Choose a style that has been around for a while and is not in danger of going out of fashion. No one likes to change container types in mid-collection.

It often comes down to a personal preference after weighing all the factors, or simply what you have on hand.

If you're just starting your sand collection, talk to other collectors and see what they're using and why they favor one style over another. If you've inherited or purchased an old collection, you may want to give some thought to re-packaging the sand for consistency's sake. Also, faded labels may need to be re-written (but try to keep the original labels for provenance).

With so many choices, it's easy to become overwhelmed. When all is said and done, it's not as important *how* you do it, as long as you begin.

References:
Author's own, twisted brain
Non-Copyrighted Photos (from the Internet) used under the Fair Use Policy

Heidi Scheirer McGrew

Sand Painting:
Sand as an Artistic and Religious Symbol

Sand painting is one of the oldest art forms: it has been practiced and continues to be practiced by Buddhist monks in Tibet; Native Americans in the Southwestern U.S.; Aboriginals in Australia; and native peoples in Africa, Asia and Polynesia. Modern Latin Americans sometimes practice it during the "Dia de los Muertos" (the Day of the Dead). Traditionally, sand painting has been used for religious ceremonies, often associated with healing. By its nature and intent, these traditional sandpaintings have been temporary, highly symbolic and sometimes secretive.

Perhaps the most well-known sand painting is the Tibetan tantric Buddhist tradition of creating elaborate mandalas (known as dul-tson-kyil-khor). Made from colorful sand, these intricate constructions are created on the ground or on a low platform by a team of monks. The design is both geometric and symmetric and includes religious symbolism. The creation of these sand paintings is accompanied by religious rituals, blessings, meditations, chants, etc. Traditionally done for healing of individuals, modern mandalas are sometimes constructed for the wider purposes of healing the environment or the Earth.

Large mandalas take several days or even weeks to construct. An essential aspect of the ceremony is that a mandala is destroyed soon after it is completed. They are constructed from the center outward and when they are destroyed, the sand is swept from the outer edges in to the center. So they both begin and end at the center. The impermanence of the design represents the impermanence of life.

Another well-known tradition comes from the American Southwest where Native Americans have constructed sand paintings for thousands of years. The Navaho are particularly associated with sand painting although the tradition is widespread among the Southwestern tribes. Highly symbolic, these sand paintings are constructed on the ground or a low platform by medicine men as part of their many different healing ceremonies. Believed to be actual representations of the heavens and creation, individual sand paintings are specific for each healing ceremony.

It is very important that the sand painting be created exactly the right way each time since it is believed that the gods temporarily inhabit the representations during the height of the ceremony. At the end, the sand paintings must be destroyed in order to release the gods. Apprentice medicine men spend years learning the rituals and the exact designs of the sandpaintings for each ceremony.

Beginning in the mid-20th century, some Native Americans began making permanent representations of sand paintings using traditional designs from the healing ceremonies; however, they deliberately altered the designs so as not to anger the gods who might think that humans were trying to trap them. It is thought that most of the permanent Native American sand paintings include these deliberate errors or omissions.

For thousands of years, native Aboriginals in Australia have used earth materials such as sand, stones, and sticks to create designs on the ground as part of their religious ceremonies. Like the Native American sand paintings, these designs are symbolic, specific to a ceremony and highly secretive. Men's and women's ceremonies have different designs and the knowledge or even the sight of these designs is forbidden to the opposite sex. In order to keep the designs secret, they are destroyed as soon as the ceremony is completed. It is thought that many of these traditional on-the-ground constructions give specific information about travel and journeys as well as creation stories.

About 1950, a few aboriginal artists began to make representations of these ritual sand paintings in what became known as "dot paintings." Constructed of small circles ("dots"), these paintings are usually done in acrylic paints and are colorful and dramatic. The subjects include images from aboriginal ceremonies, but, like the Navaho sand paintings, dot paintings are created with deliberate omissions and alterations. Dot paintings have become quite popular; they are featured in many art galleries; and some sell for thousands of dollars.

A contemporary sand painting tradition comes from Latin America and the Southwestern U.S. where sand paintings are associated with the Day of the Dead ("Dia de los Muertos"). Impermanence is also an important aspect of these paintings. Many African, Asian and Polynesian societies have sand painting traditions. A common thread among these is the use of symbolism, religious ritual, and impermanence. Several of these religious traditions have evolved into more cultural and artistic applications.

Contemporary sand paintings depicting realistic subjects are created using a variety of techniques in such varied places as Bali, Senegal, Nigeria, China and Japan.

Sand painting has been an almost universal art form in human societies. Often associated with healing ceremonies and creation stories, religious sand paintings have many things in common: they are usually created on the ground; they are constructed with sand and other Earth materials; and they are deliberately impermanent. Some of these religious sand painting traditions have been adapted into modern art forms intended to be permanent. In addition, contemporary sand artists have been inspired by the beauty and color of sand to create representations of their world.

References

Caruana, Wally, *Aboriginal Art, 2nd edition*, 2003, Thames and Hudson Publishers, 232 pp.

London, Eileen and Belinda Recio, *Sacred Rituals: Connecting with the Spirit through Labyrinths, Sand Paintings and Other Traditional Arts*, 2004, Fair Winds Press, Gloucester, MA, 192 pp.

McCoy, Roy, *Summoning the Gods: Sandpainting in the Native American Southwest*, 1988, *Plateau Magazine, v. 59, no. 1*, Museum of Northern Arizona, Flagstaff, AZ, 32 pp.

Pike, Brian, Wikipedia sand painting entry, http://en.wikipedia.org/wiki/Sandpainting

Villasenor, David, *Tapestries in Sand: the Spirit of Indian Sandpainting*, 1963, Naturegraph Publishers, Happy Camp, CA, 112 pp.

Image Credits

Tibetan Sand Mandala, Olga Jarrett, 2007, Carlos Museum at Emory University, Georgia

"Heading People," Navajo ceremonial sand painting, Angelina Myerson (age 11), 6"x12", no date.

"Women's Ceremony," Beatrice Napurulla, 13"x19", canvas, no date

Woman grinding grain with baby, 9"x11", African, no information about author or date. Even the frame is sand painted!

Boat and setting Sun, 10"x16", Dakar, Senegal. Initials: BH or BS or BSH. "CaurJ" signed on the back. Also Baobab Galerie Africaine with address.

Chinese pavilion, Chinese sand and crushed stone painting, 10"x14", No information about author or date, Back says "Weihao" which may be the name of the company that made the board.

Heidi Scheirer McGrew

A rock collector since childhood and a geophysicist by training, Heidi has taught general earth science classes at the University of Dayton for 17 years. She began collecting sand on a field trip to New Zealand; returning home, Heidi experimented until she taught herself to how to sand paint. She feels fortunate to be able to practice what is perhaps the world's oldest art form: a craft which allows her to combine art with her love of and fascination with nature.

You can reach her at naturaleartcraft@gmail.com
Her sand painting blog is at:
http://naturalearthcraft.blogspot.com/
She sells her sand paintings and stone jewelry on Etsy at:
https://www.etsy.com/shop/NaturalEarthCraft

Lori Carter

Heavy Sand Processing:
How Heavy Minerals Are Extracted and Separated

This presentation was compiled from a Georgia Mineral Society field trip to the DuPont Florida Plant and from information provided by Iluka Resources. The heavy sand photograph is by Kate Clover.

What is Heavy Sand?

- Sand sized particles of minerals with specific gravity heavier than quartz, which is usually the main component in surrounding minerals
- They are often found in natural accumulations of heavy minerals called ***placer deposits***
- Separation occurs during sedimentation processes as eroded materials are moved by water, wind, and gravity

What is the definition of “sand sized”? There are technical specifications for example 1/6mm to 2mm, but for a sand collector, it seems if you want to call it sand then it is sand!

Examples of Heavy Sand Minerals

- Rutile – TiO_2
- Ilmenite – $FeTiO_3$
- Zircon – $ZrSiO_4$
- Staurolite – $Fe^{2+}{}_2Al_9O_6(SiO_4)_4(O,OH)_2$
- Rare Earth Minerals
- Magnetite – Fe_3O_4
- Garnet - $(Fe,Mn,Ca)_3.Al2(SiO4)_3$

There are other heavy minerals in heavy sand, but these are some of the more common minerals. The first 5 listed have commercial, economic value, though magnetite and garnet have value in other non-commercial and fun ways.

Importance of Heavy Sands

- Industrial and Commercial Uses – as an opacifyer in paint, paper, plastics; in high temperature applications e.g. metal casting and ceramics; and as abrasives
- They are a good indication of the presence of gold so they are often referred to by gold seekers as "black sand"
- Fun to look at and study – the mineralogy is varied and interesting

Even though the heavy minerals mined often comprise less than 10% of the sand extracted, heavy sand mining can be a lucrative business.

How Rutile is Used

Rutile (titanium dioxide) is used in titanium metal alloys and as a white pigment in paper, paint, plastics, medicine, toothpaste – even the little m's on M&M's!

M&M's is a registered trademark by Mars, Inc.

Photo is by Evan-Amos via Wikimedia Commons

Titanium metal alloys can be used for jewelry, joint replacement, screws and pins to hold fractured bones in place, bicycle frames and other sports equipment, in aerospace applications – anywhere strong but light weight, non-reactive, and non-ferromagnetic materials are needed.

How Zircon is Used

Zircon (zirconium silicate) is used for high temperature applications like bricks in steel and glass industries and as an opacifier in ceramics.

Pouring sand castings at the Ford Foundry in Dearborn, MI

Photographer: Joe Clark

National Archives and Records Administration

Zircon (zirconium silicate) is chemically inert and stable at high temperatures. It has a melting point of 3992°F (2200°C) [Iluka Resources]

How Staurolite is Used

Staurolite sand is often used as an abrasive for sand-blasting stone.

Sand-blasting in progress at a granite monument company

Photo by Lori Carter

A rubber template with the desired wording/design is adhered to the stone, then the staurolite sand is sprayed through a high pressure hose. The sand etches the stone where it is not protected by the template.

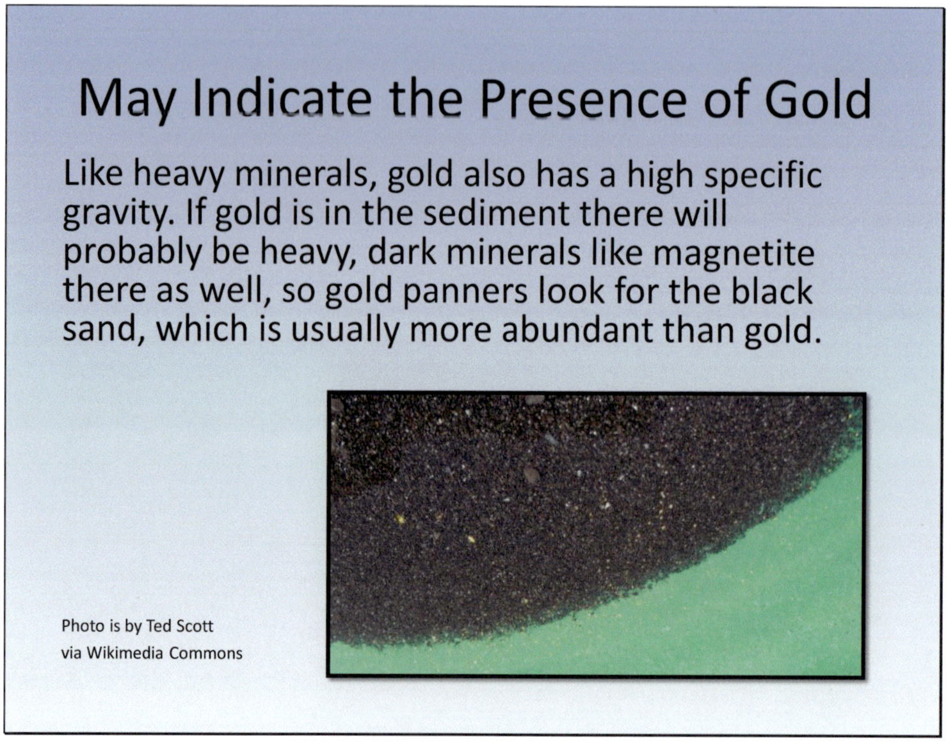

Gold panners often save the black sand they find as concentrate, then look for the gold later.

This example of heavy sand includes ilmenite (black), staurolite (red and orange), and zircon (rounded clear and white). It also contains quartz (angular clear and white). This sand was mined from an area that was once a shoreline characterized by a beach ridge with barrier islands and inland dunes.

Processing Steps

1. Dredging (or Mining)
2. Wet Mill Separation
3. Dry Mill Separation
4. Reclamation

DuPont Florida Plant image from Google Maps

A Google Maps aerial view of the DuPont Florida Plant shows a main dredge and a scavenger dredge with pipes connected to the floating wet mill. Tailings are ejected behind the wet mill.

(1) Dredging

A slurry of sand and water is dredged from a pond

DuPont Florida Plant photo by Lori Carter

The dredge uses a "cutter head", which is like a large drill bit, to break up the edges of the pond so the sand can be sucked up. Large pieces are screened and ejected before the slurry of sand and water is sent through a pipe to the wet mill.

(2) Wet Mill

Some of the wet mill separation (spirals) can be accomplished in the same pond where dredging is performed. Other wet mill processes are done at a wet concentration plant. Wet mill separation may include, but is not limited to:

- Spirals
- Up-Current Classifier
- Wet Table

Wet mill processing can eliminate 80 to 90 percent of the unwanted materials in just the first step (spirals).

Wet Mill Spirals

- Spiral separators use gravity and centrifugal force to separate heavy sand from lighter materials
- The lighter materials are flung to the edges of the spiral and are eventually ejected as "tailings"
- The heavier minerals stay to the inside of the spiral and go down a trough where they are extracted as heavy sand concentrate

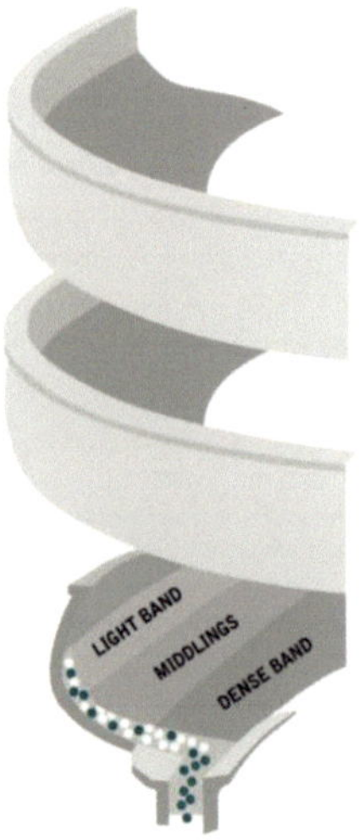

Diagram courtesy of Iluka Resources

The spirals work like an amusement park ride I rode with my brother-in-law who outweighed me by at least 2 to 1. We were in a large cylinder that rotated faster and faster until centrifugal force moved us up the sides of the cylinder. At one point I shot way up the wall but my brother-in-law remained at the bottom. If we were in a wet mill spiral, he would be the heavy sand and become part of the concentrate. I would become part of the tailings and be ejected.

Up-Current Classifiers

Upward currents of water float lighter material away while heavier material sinks and is funneled down through the conical portion of the classifier

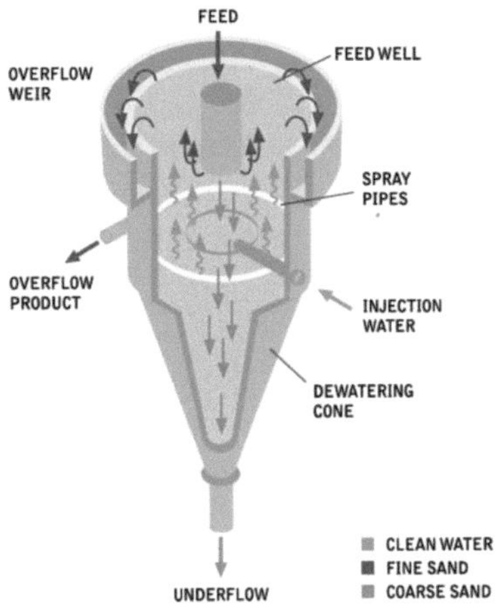

Diagram courtesy of Iluka Resources

Sand collectors can use a process similar to this. Pour sand into a large container of water, swirl it around, then pour the water out. This is a very good way to remove fine particles like clay as well as organic materials like seeds, leaves, and twigs.

Wet Table Separator

- Wet tables have a rippled surface with water flowing perpendicular to the ripples
- Lighter grains are carried by the water past the ripples while heavier particles cannot be moved over the ripples

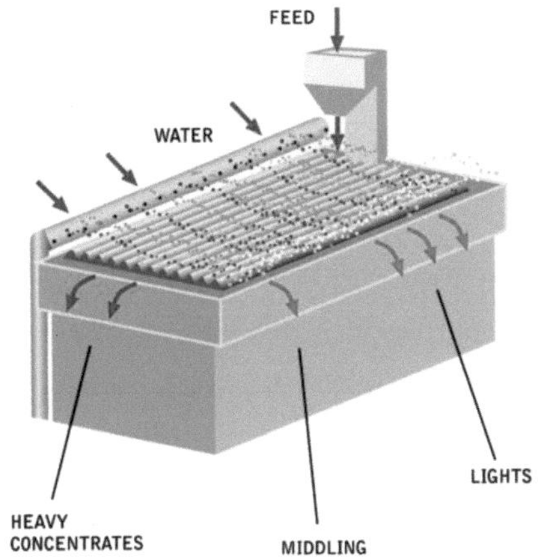

Diagram courtesy of Iluka Resources

This process is similar to the way a gold pan works. A gold pan has ridges in the bottom on one side. When water is swirled around the pan the lighter sand can be floated away while the heavier sand (and gold) is caught by the ridges.

(3) Dry Mill

The Dry Mill uses various techniques to separate individual minerals from the Wet Mill concentrate including, but not limited to:

- Electrostatic Separator
- Magnetic Separator
- Air Table

Other processes may be employed at the dry mill including screening and additional spiral processing.

Electrostatic Separator

- An ionizing wire electrode charges sand grains
- Conductive minerals quickly lose their charge to the grounded rotating roll and are flung away by centrifugal force as well as being attracted to another electrode plate
- Non-conductive minerals have a charge opposite to the rotating roll, so they have a strong attraction and remain on the roll until brushes remove them

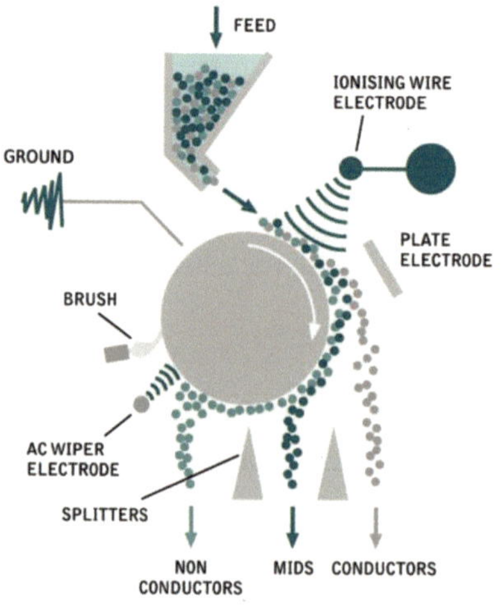

Diagram courtesy of Iluka Resources

The "mids" in this diagram are the minerals that fall between conductive and non-conductive. They are not necessarily tailings or rejects so they may be passed through additional separation processes.

Magnetic Separator

- Materials that are not attracted to the magnetic roller fall off the moving belt
- Minerals that are attracted to the magnetic roller remain on the moving belt until they leave the influence of the magnetic field and drop off

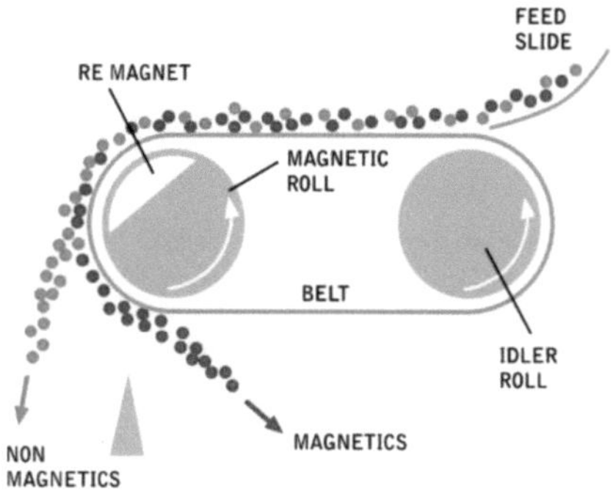

Diagram courtesy of Iluka Resources

Most of the ilmenite can be extracted via magnetic separation, but weathered and altered ilmenite loses Fe and becomes less magnetic, so electrostatic separation is used to extract residual ilmenite (Iluka Resources)

Air Table Separator

- Sand is fed onto an angled vibrating surface covered with a woven cloth
- As air is blown up through the cloth, heavier minerals sink through the cloth and are directed one way while lighter materials rise to the top and are directed another

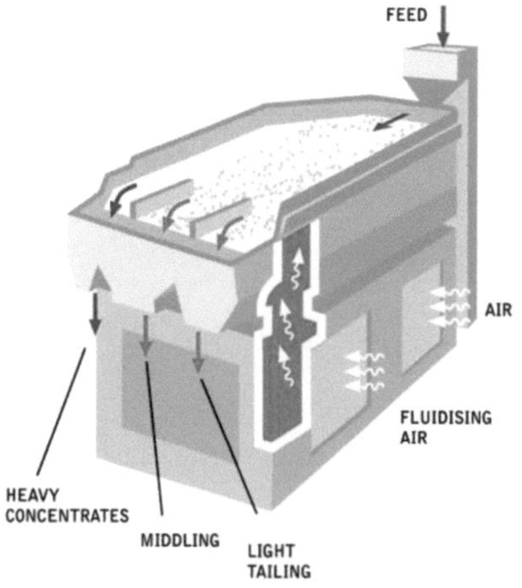

Diagram courtesy of Iluka Resources

High volume, low pressure air blows light material through the mesh cloth while heavy material sinks through the mesh and onto the angled surface of the table. The reciprocating motion of the table makes the heavier materials move up the higher side of the table and it makes the lighter materials drift to the lower side.

(4) Final Phase - Reclamation

Before mining begins, overburden containing native seeds is removed and saved. As mining proceeds, tailings are used to contour the area. When mining is done, overburden is returned and animals may be reintroduced.

DuPont Florida Plant
photo by Lori Carter

Heavy sand mining operations go to great lengths to preserve the ecology of an area being mined. Sometimes the pre-mining phase can take longer than the mining itself!

Thom Hopen, Nick D'Errico, Ralf Milke

Writing Sand (aka Blotting Sand)

This is a study in progress that began as a "curiosity question" and has turned into one of those wonderful and interesting projects. Our original goal was to look at a few samples and determine the composition of "Blotting Sand" (also commonly referred to as "Writing Sand", "Shining Sand", and "Black Sand"). The purpose of blotting sand was to quickly dry the wet ink that had been applied to sized paper so the ink would not smear. It was sprinkled on the paper by means of a small container called a "Sand Box", "Sander", "Sand Shaker" or just "Shaker". After applying the sand and the ink had dried the excess sand was poured back into the small vessel to be used again.

The use of pounce will also be briefly discussed which preceded the use of blotting sand. Pounce is a fine powder that was rubbed on to parchment and/or unsized paper so the ink would not run when applied. Pounce is reported to be pumice, powdered cuttlefish bone, and gum sandarac. It has also been reported that various ground egg shell mixtures can be used as a substitute for pounce.

The investigation started when our literature research failed to conclusively determine the make-up of blotting sand. It has been reported that biotite, ilmenite, gold, and white sand from Calais, France has been used for blotting sand. Authentic samples of pounce and blotting sands have been acquired from the original packages or containers. Residues from sand boxes have also been obtained as well as sand samples collected from documents (or the packages that the old documents have been stored in) from as early as 1553. Over 100 samples from three countries have been examined and analyzed by stereomicroscopy, PLM and SEM-EDS. The results obtained thus far will be presented and I think one will be surprised as to what we have discovered (or not discovered) in our studies. The use of pounce will also be briefly discussed which preceded the use of blotting sand. Pounce is a fine powder that was rubbed on to parchment and/or unsized paper so the ink would not run when applied. Pounce is reported to be pumice, powdered cuttlefish bone, and gum sandarac. It has also been reported that various ground egg shell mixtures can be used as a substitute for pounce.

Blotting Sand

Calis Sand
Shining Sand
Gold Writing Sand
Black Writing Sand
Round Writing Sand
Superior Writing Sand

Pounce and Blotting Sand

1780 State Invoice

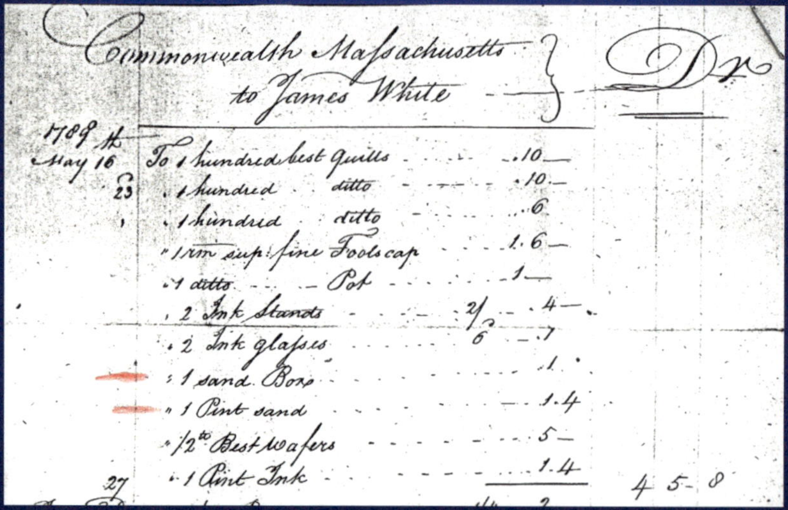
Commonwealth Massachusetts to James White Dr

1789			£	
May 16	To 1 hundred best quills		.10	
23	1 hundred ditto		.10	
	1 hundred ditto		.6	
	1 rm sup: fine Foolscap		1.6	
	1 ditto Pot		1	
	2 Ink Stands	2/	.4	
	2 Ink glasses	6	.7	
	1 sand Box		.1	
	1 Pint sand		1.4	
	1/2 lb Best Wafers		5	
27	1 Pint Ink		1.4	4 5 8

Pounce and Blotting Sand

Calis Sand from Calais, France

Pounce and Blotting Sand

Quartz Blotting Sand

Pounce and Blotting Sand

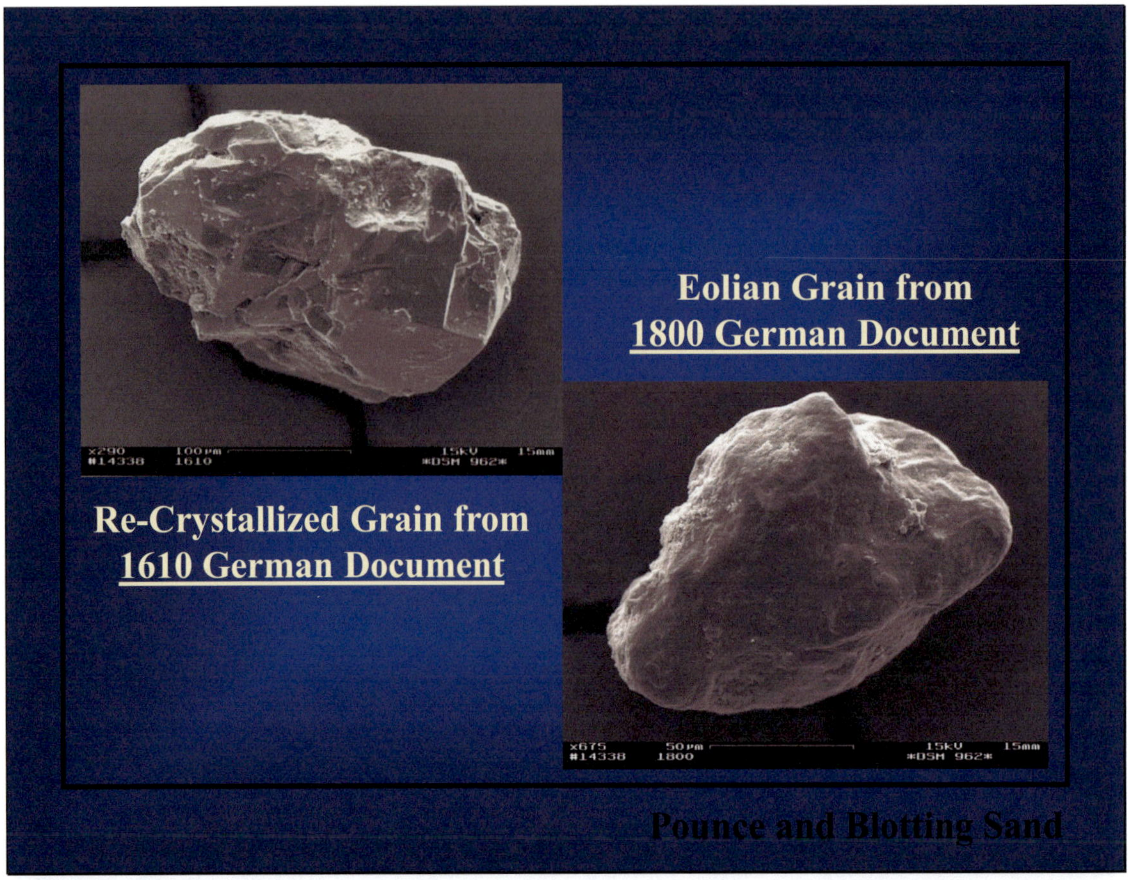
x290 100 µm 15kV 15mm
#14338 1610 *DSM 962*
Eolian Grain from
1800 German Document
Re-Crystallized Grain from
1610 German Document
x675 50 µm 15kV 15mm
#14338 1800 *DSM 962*
Pounce and Blotting Sand

OGDEN'S CIGARETTES
A BATTLE ON THE SANDS: ALEXANDRIA, 1801
THE STORY OF SAND
A SERIES OF 50
(Sand in History).
4
A BATTLE ON THE SANDS : ALEXANDRIA, 1801.
In March, 1801, a British expeditionary corps under Lt.-Gen. Sir Ralph Abercromby was despatched to Egypt to expel or capture the French Army left there by Napoleon. In the face of strenuous opposition the British landed near Aboukir (at the eastern end of the sandy spit of land on which Alexandria stands), a daring and brilliant exploit. We illustrate an episode of the landing among the sand-hills, when Col. Brent-Spencer of the 40th Regiment, one of the first ashore, displayed great gallantry. At the Battle of Alexandria the French were defeated, though Abercromby was wounded by a spent ball and died a week after the victory.
ISSUED BY
· OGDEN'S ·
BRANCH OF THE IMPERIAL TOBACCO CO. (OF GREAT BRITAIN & IRELAND), LTD.
OGDEN'S CIGARETTES
SAND FOR DRYING WRITING-INK
THE STORY OF SAND
A SERIES OF 50
(Sand in History).
6
SAND FOR DRYING WRITING-INK.
Although blotting-paper appears to have been known as long ago as 1465 its use did not become general until about the middle of last century. During the 18th and early 19th centuries fine white sand was still widely used for drying handwriting. A writer in All the Year Round (1860) describes one of his characters as "continually shaking sand from a pepper-box over scrawling entries in marble-covered copy-books." Old inkstands of china, silver, or Sheffield plate were sometimes fitted with sand-boxes; the specimen illustrated is of English china and dates from about 1815.
ISSUED BY
· OGDEN'S ·
BRANCH OF THE IMPERIAL TOBACCO CO. (OF GREAT BRITAIN & IRELAND), LTD.
Pounce and Blotting Sand

SUPERIOR
BLOTTING SAND.
Lake Champlain.
FOR SALE
BY ALL
ROUND STYLE
SAND.
Pounce and Blotting Sand

Black
Blotting Sand
OSV-2
500 µm
OSV-2
300 µm
Pounce and Blotting Sand

Blotting Sand Vessels

Sand Boxes
Sand Shakers
Sanders

Pounce and Blotting Sand

Glass Sander

Pounce and Blotting Sand

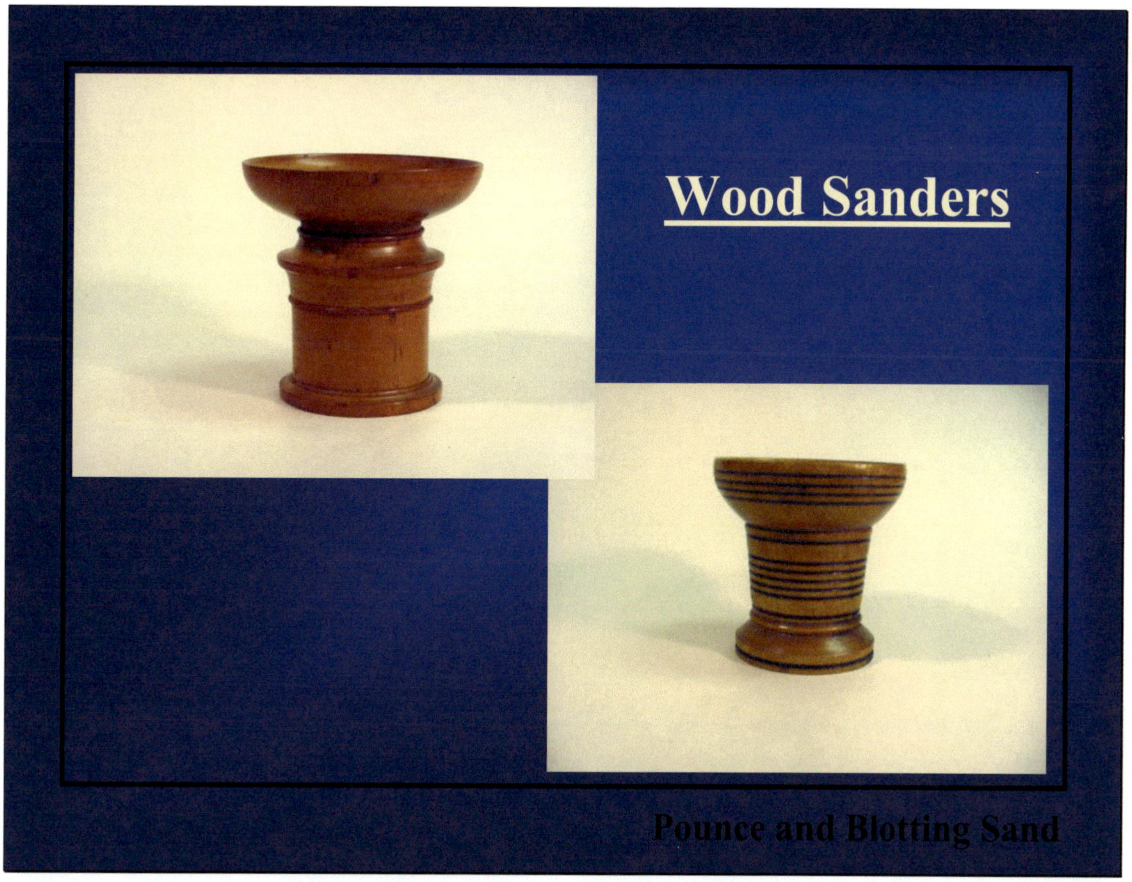
Wood Sanders
Pounce and Blotting Sand

Metal Sander
Wood Sander
Pounce and Blotting Sand

Pounce and Blotting Sand

Blotting Sand
from Sander
Pounce and Blotting Sand

Sand from
John Hancock
Document 1784
Pounce and Blotting Sand

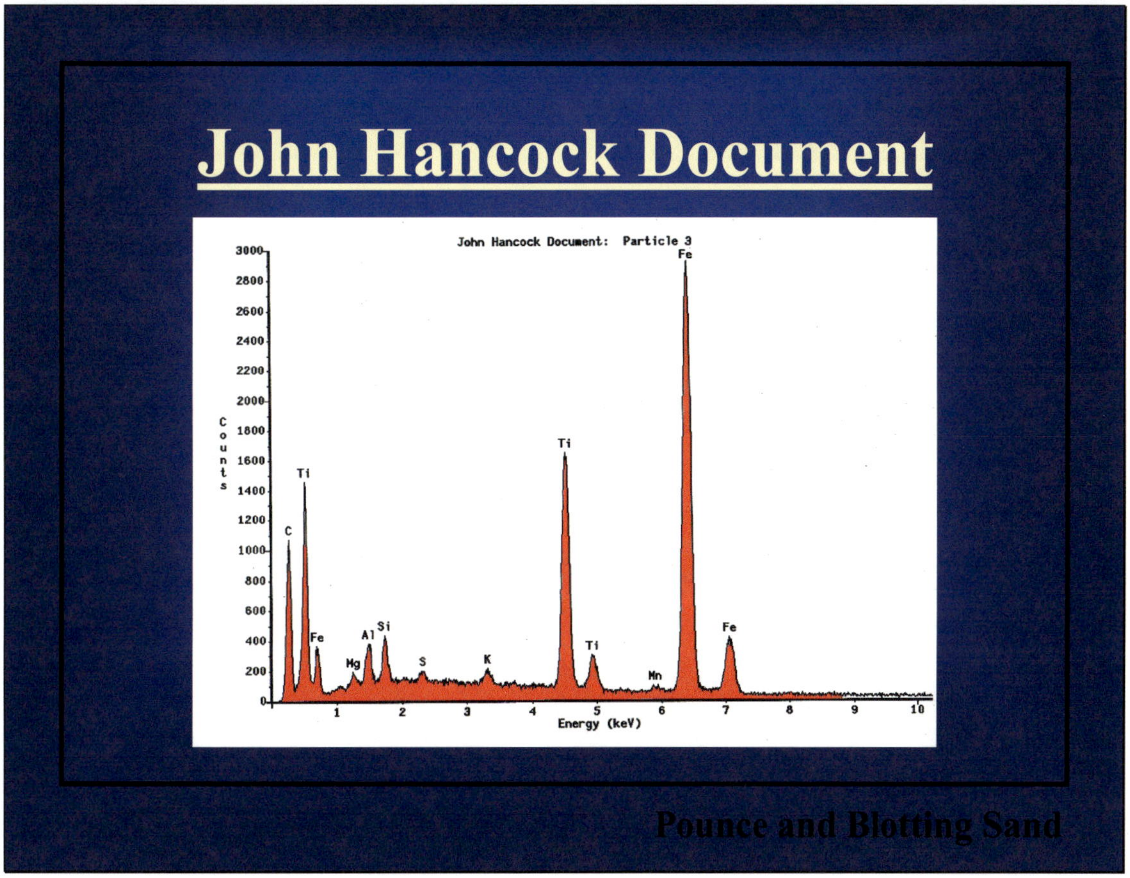
John Hancock Document
John Hancock Document: Particle 3
Counts
Energy (keV)
C
Ti
Fe
Mg
Al
Si
S
K
Ti
Ti
Mn
Fe
Fe
Pounce and Blotting Sand

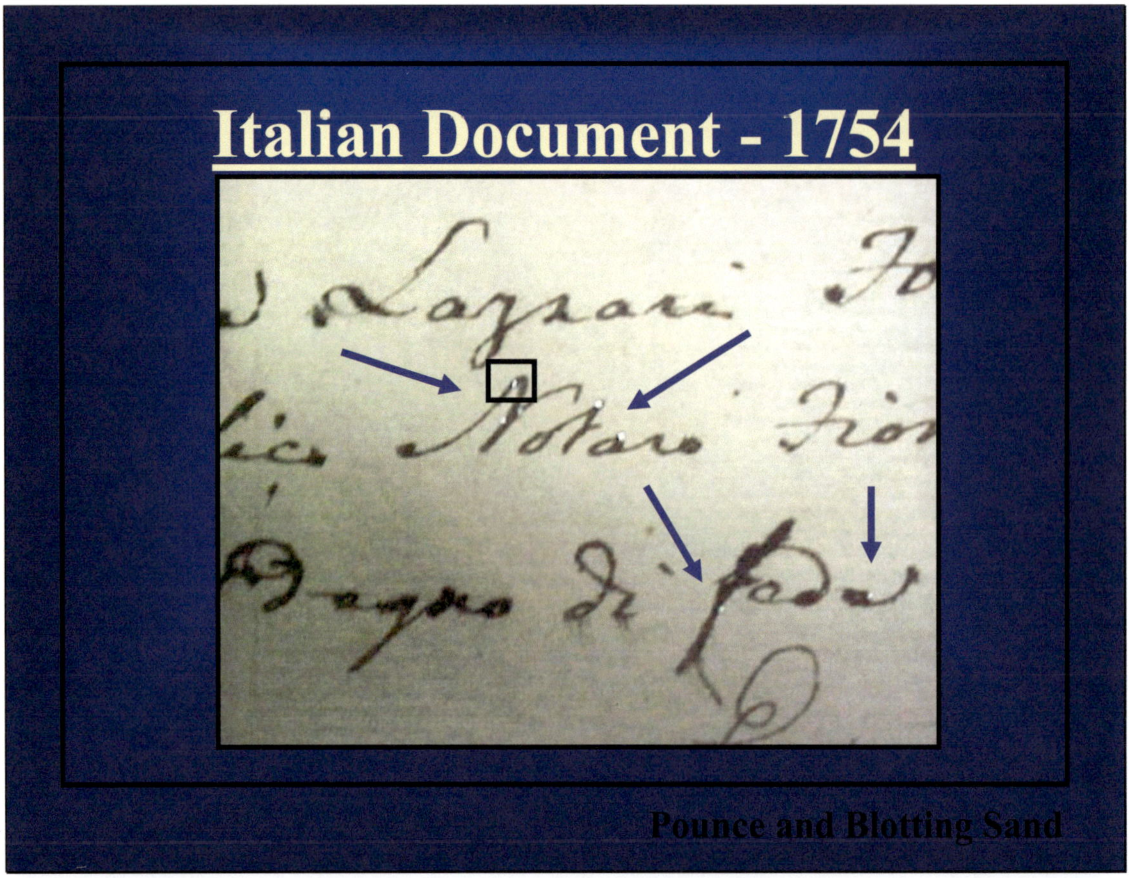
Italian Document - 1754
Pounce and Blotting Sand

Italian Document - 1754
Pounce and Blotting Sand

Italian Document - 1754

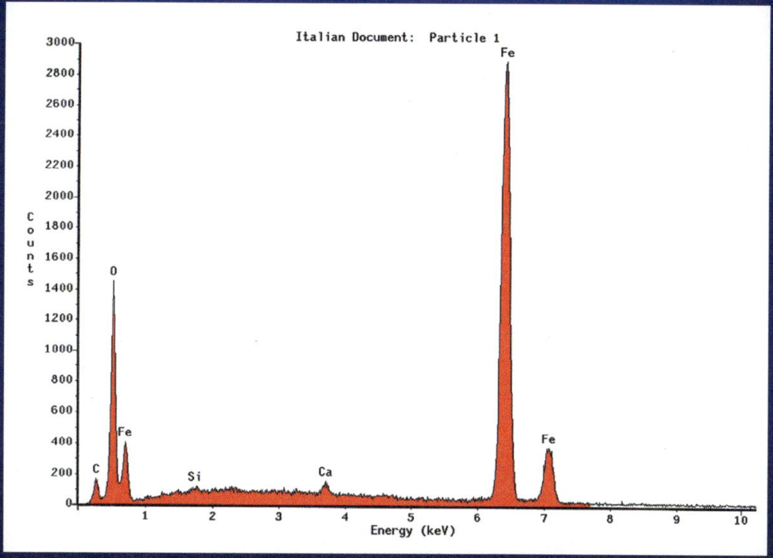

Pounce and Blotting Sand

Why Sand

Density (?)
Cosmetic (?)
Black Sand Looked Like Ink (?)
Did It Really Work (?)

Pounce and Blotting Sand

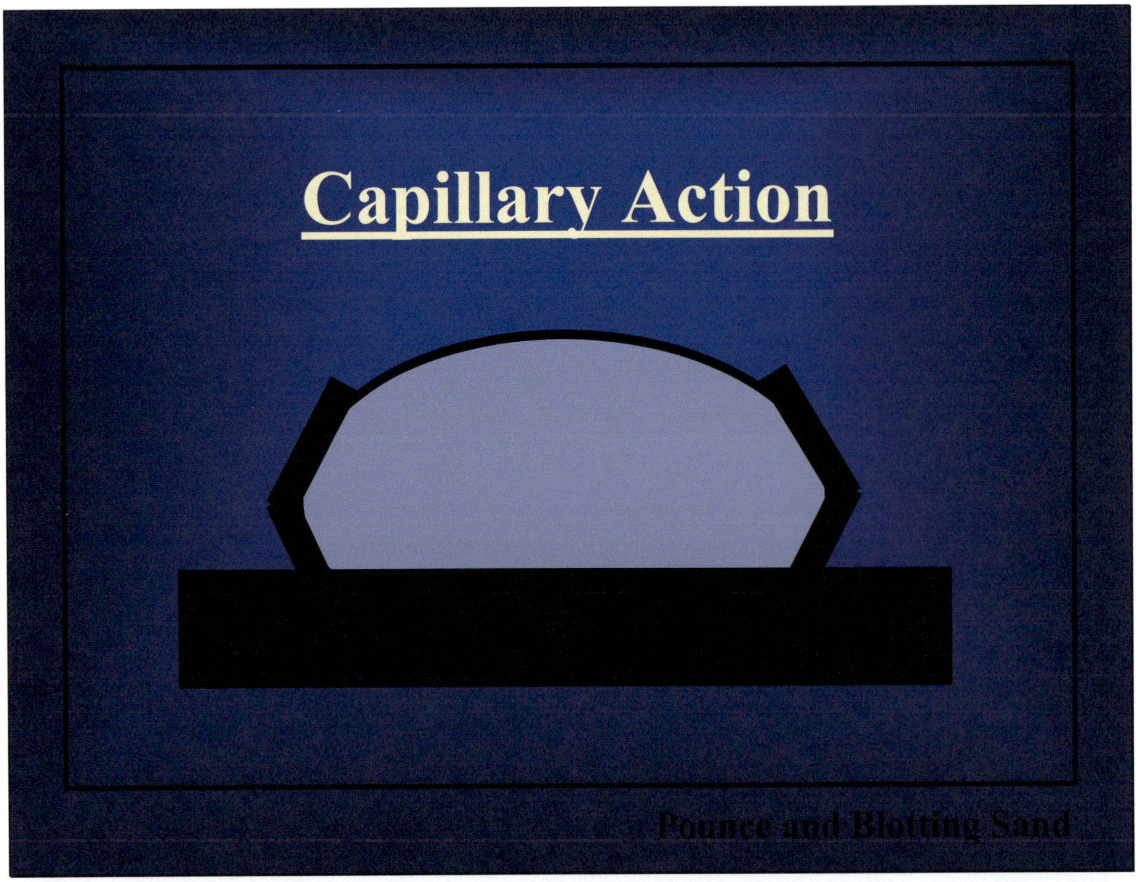
Capillary Action
Pounce and Blotting Sand

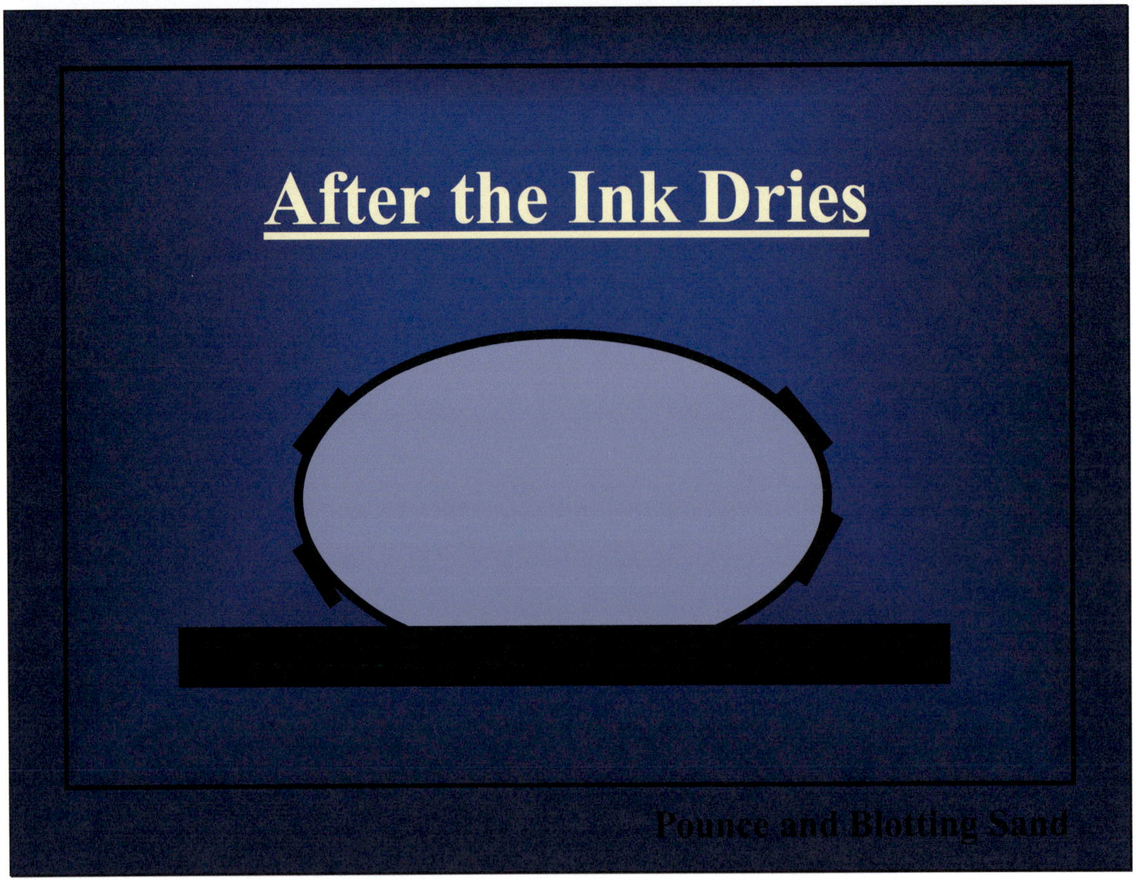
After the Ink Dries
Pounce and Blotting Sand

Grain Size Range

100 μm to 300 μm
Some as small as 75 μm

Pounce and Blotting Sand

Surface Area

Decrease Grain Size by ½
Increase # of Grains by 8 in Sample
Single Grain Surface Area Decreases by ¼
Volume Constant in Sample
Surface Area of Sample Increases by 2

Pounce and Blotting Sand

Brass Particles

Pounce and Blotting Sand

Not True

Blotting Sand Was Used Because Blotting Paper Had Not Yet Been Invented

Pounce and Blotting Sand

Moby Dick by H. Melville

"Nantucket! Take your map and look at it. See what corner of the world it occupies; how it stands there, away off shore, more lonely than the Eddystone lighthouse. Look at it - a mere hillock, and elbow of sand; all beach, without a background. There is more sand there than you would use in twenty years as a substitute for blotting paper."

Pounce and Blotting Sand

Advertisements

Both Pounce and Blotting Sand were Offered at the Same Time

Pounce and Blotting Sand

Click below for additional view(s)

Image 2 Image 3 Image 4

Item: **Chinese Export Nanking Blue and White Porcelain Sander, c. 1790**

Description: **Chinese Export Porcelain Nanking Blue and White Sander, c. 1790. This piece is also called a Pounce Pot and has typical Nanking decoration and a spear border. Perfect condition and quite rare.**

Pounce and Blotting Sand

Pounce

Used to Prepare Parchment and Unsized Paper for Writing

Pounce is Not Blotting Sand and Was Employed Prior to the Use of Blotting Sand

Pounce and Blotting Sand

Pounce Jar
Pounce and Blotting Sand

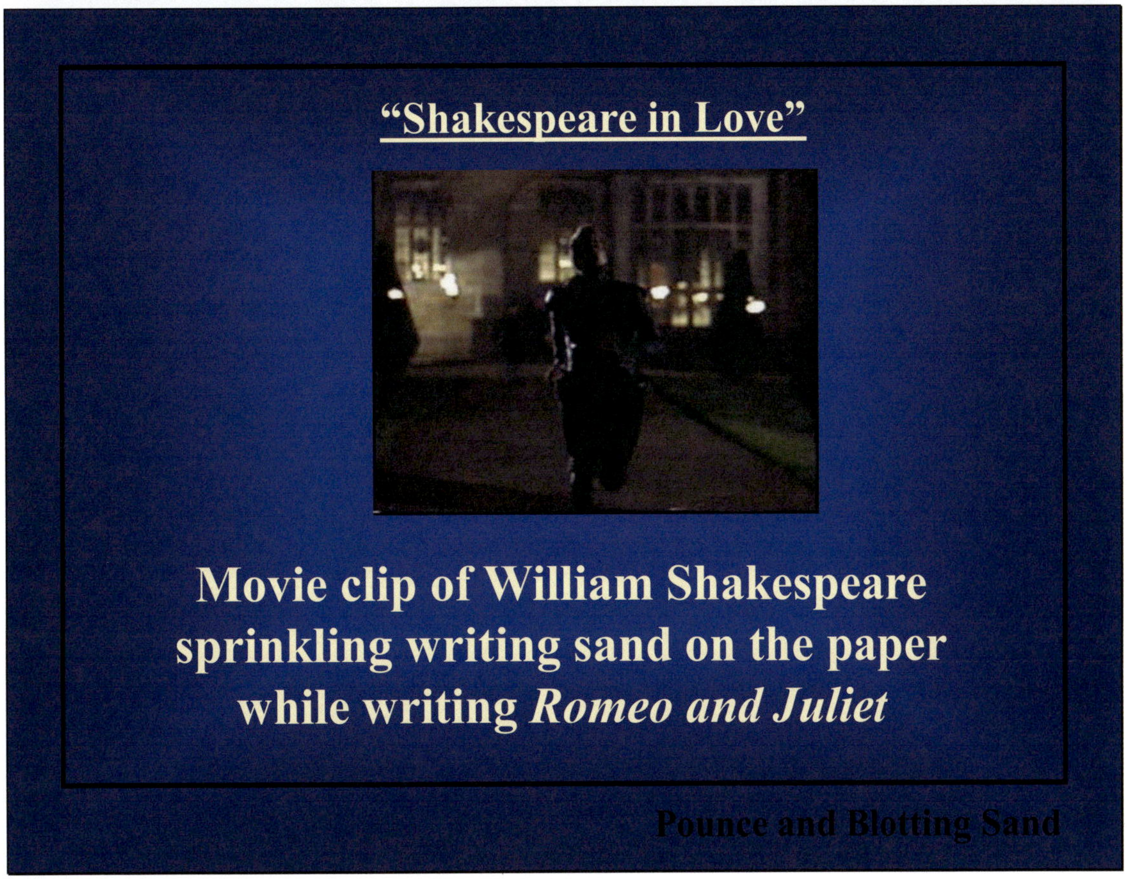
“Shakespeare in Love”
Movie clip of William Shakespeare sprinkling writing sand on the paper while writing *Romeo and Juliet*
Pounce and Blotting Sand

Robert E. Hermes, PhD

Atomic Ant Sand

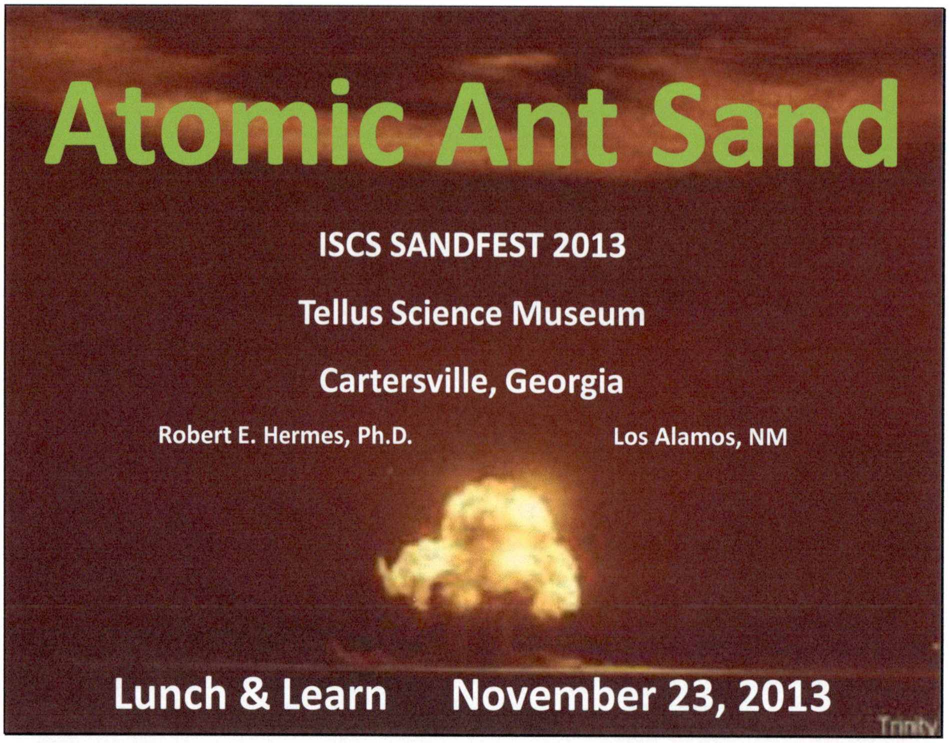

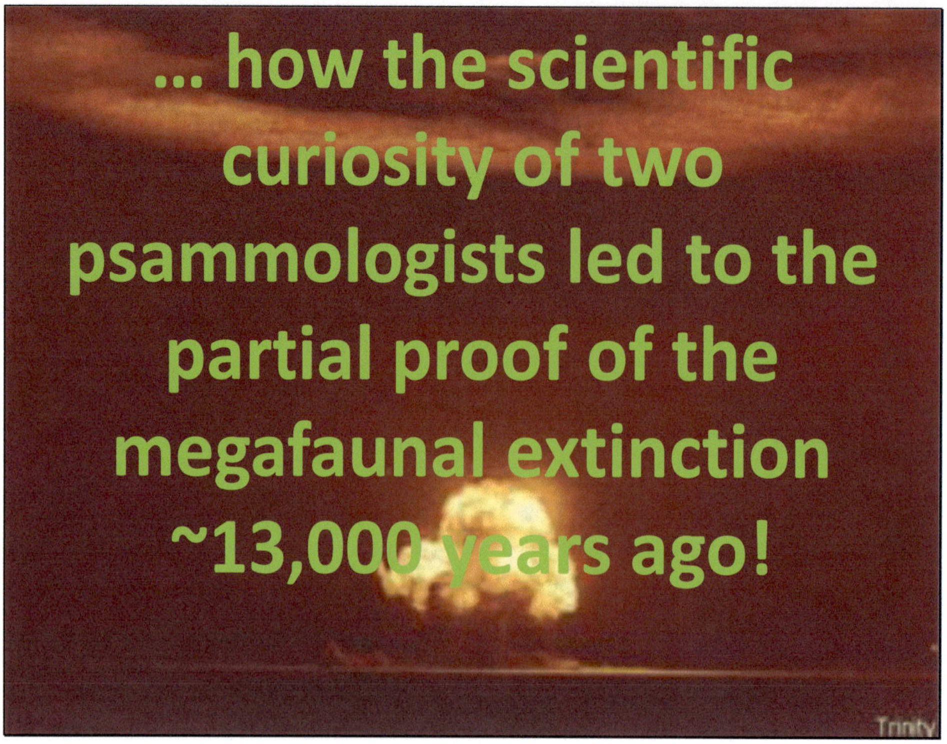

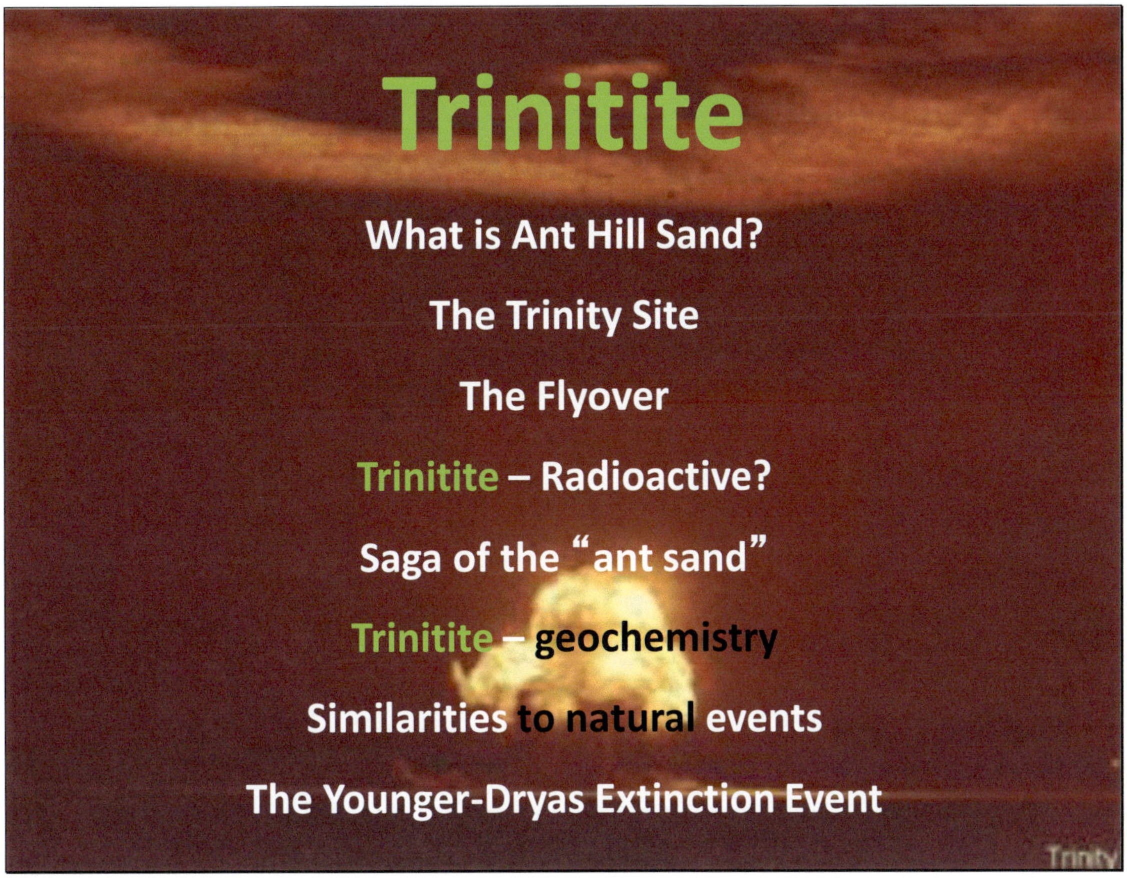
Trinitite
What is Ant Hill Sand?
The Trinity Site
The Flyover
Trinitite – Radioactive?
Saga of the "ant sand"
Trinitite – geochemistry
Similarities to natural events
The Younger-Dryas Extinction Event
Trinity

08 20 2011 12:29

08 26 2011 17:32

08 08 2011 13:35

07 19 2011

08 22 2011 19:41

The Trinity Site
TRINITY
SITE
Dr. William B. (Mr. Bill) Strickfaden at the Historical Marker for TRINITY SITE, 2004
Trinity

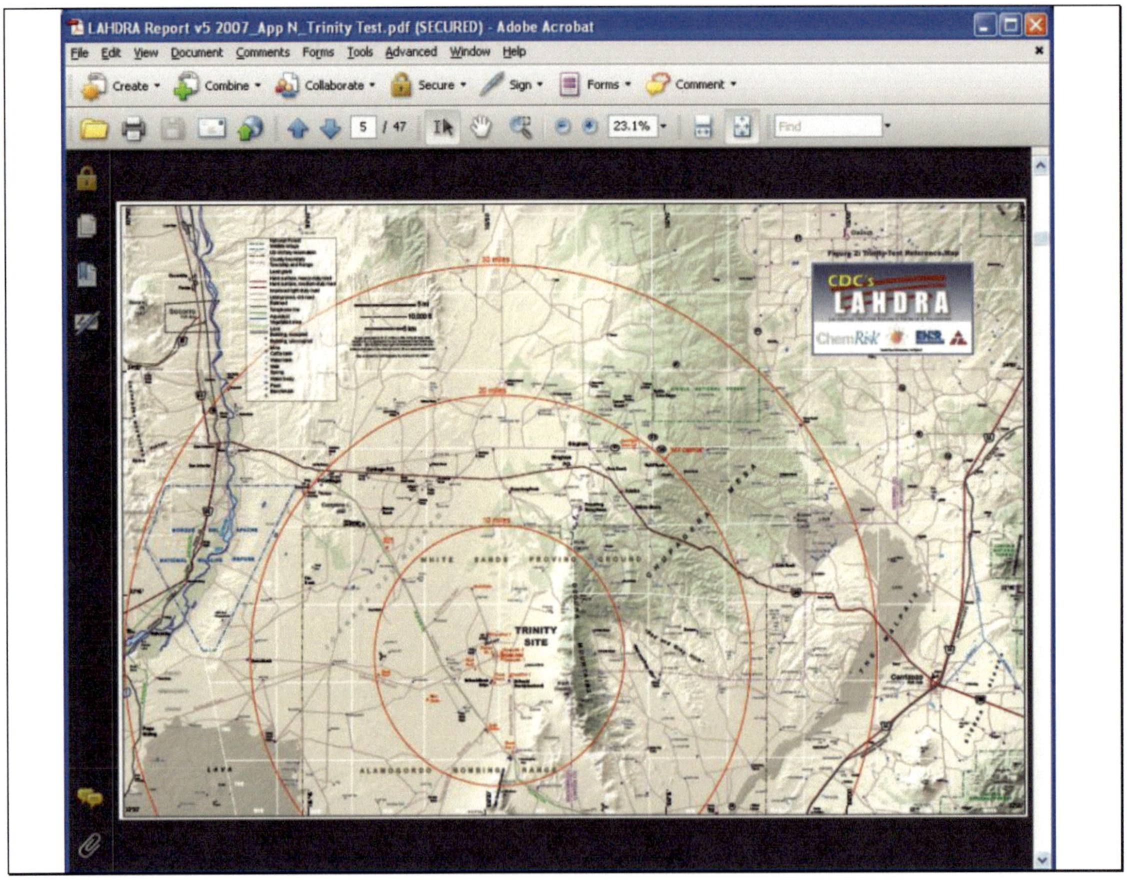
LAHDRA Report v5 2007_App N_Trinity Test.pdf (SECURED) - Adobe Acrobat
File Edit View Document Comments Forms Tools Advanced Window Help
Create Combine Collaborate Secure Sign Forms Comment
5 / 47
23.1%
Find
CDC's
LAHDRA
ChemRisk
WHITE SANDS PROVING GROUND
TRINITY
SITE
ALAMOGORDO BOMBING RANGE

Geologic map of central New Mexico

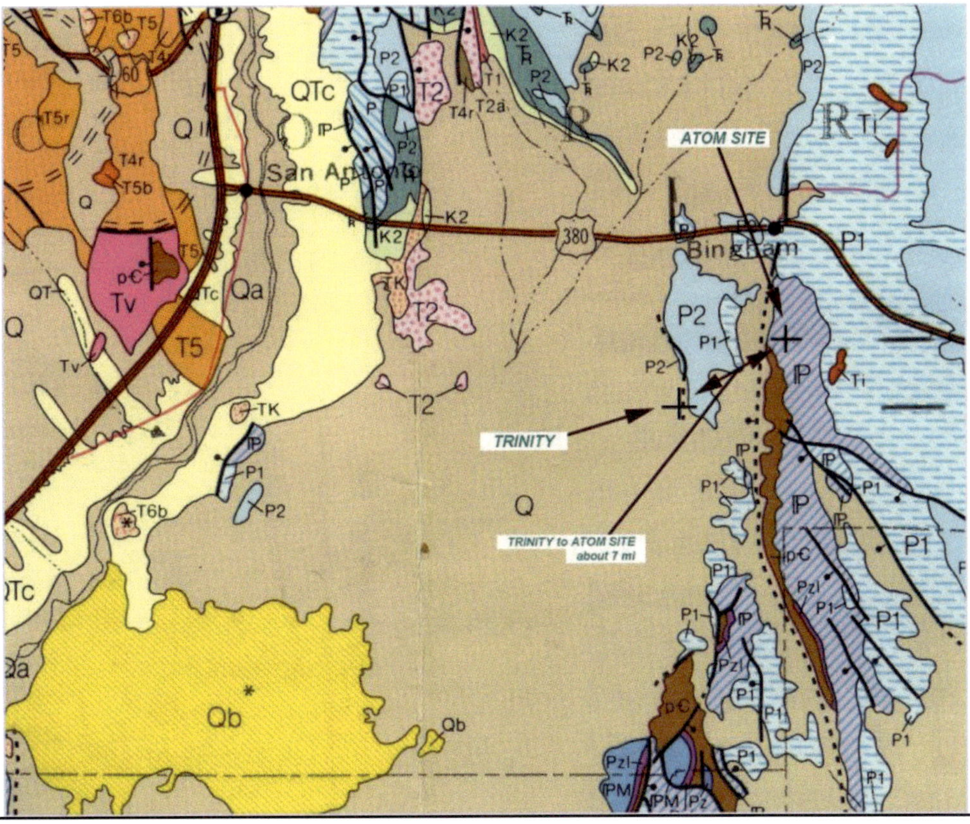

Lane, W., "New Mexico's Atom Site", *Holiday*, Nov. 1946
Coined the term ***Atomsite*** for the greenish glass

Aerial view (looking west) of the first atomic bomb's dark crater surrounded by a lighter splash created when the explosion's heat melted the sand to glass in an area encircled by a high wire fence which is patrolled by military police on horseback, in the desert.

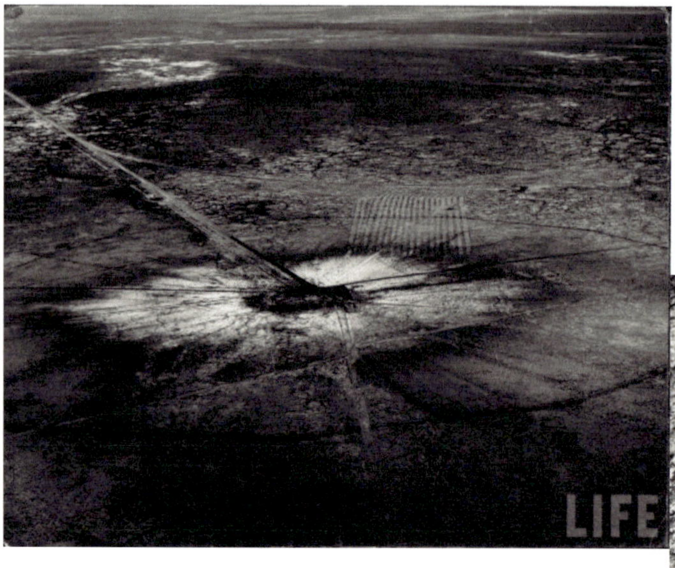

Close-up of glassy puddles

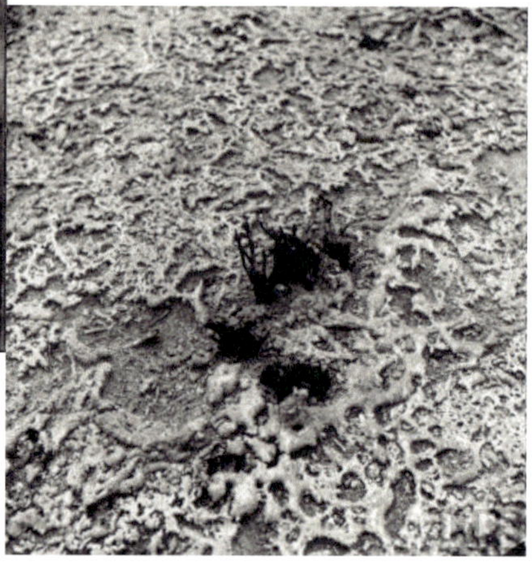

Fritz Goro, photographer, **LIFE** Magazine, Sept. 1945

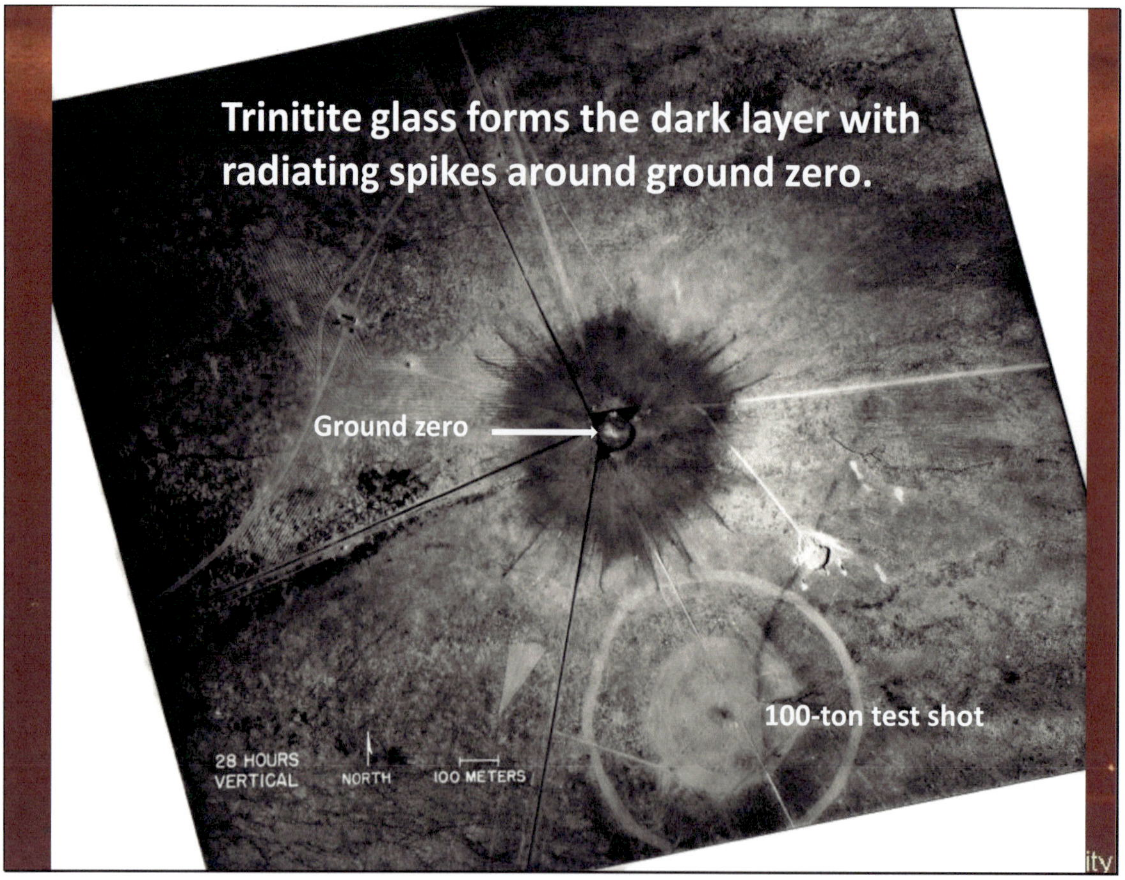

Trinity Site "today"

Composite panoramic photo by Omar Juveland, LANL Bradbury Science Museum

R. Hermes photo

R. Hermes photo

Jim Eckles – WSMR (retired)

Lead Brick Shielding

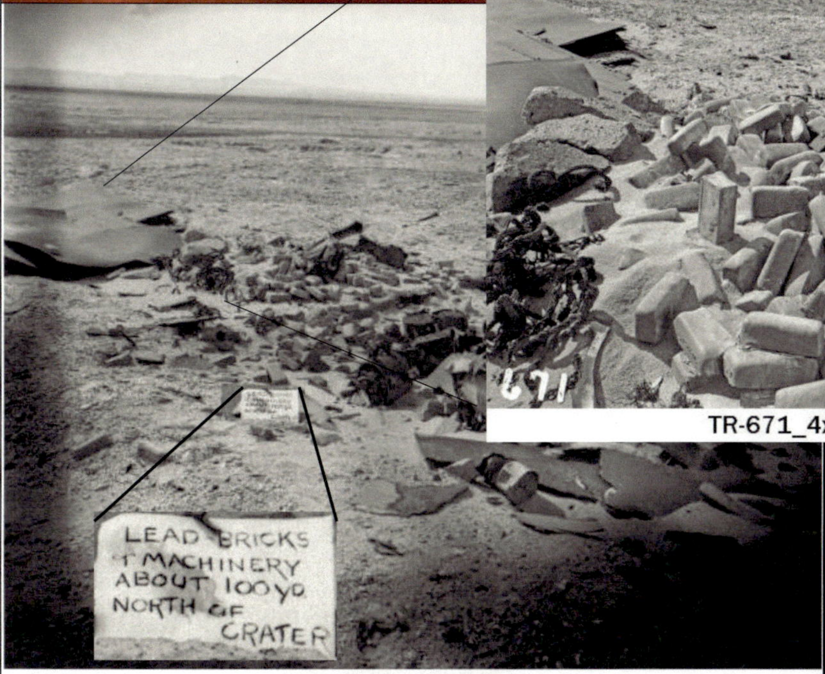

TR-662_4x5.jpg

TR-671_4x5.jpg

LANL (LASL) archive photos

Trinity

Kindly donated by
Robert Gray, ABQ, NM

FIRST ATOMIC BOMB
FROM TRINITY - SITE Y
SOCORRO COUNTY, N.M.
16 JULY 1945
GIVEN TO SHERIFF BAINARD GRAY
BY DAVE MCDONALD

"Sheriff Bainard Gray was my grandfather and Sheriff of Socorro County at the time of the test and the McDonald ranch is where the test was conducted with the McDonald ranch house used for final bomb assembly."

R. Gray

Melted lead, with Trinitite droplets fused into the surface

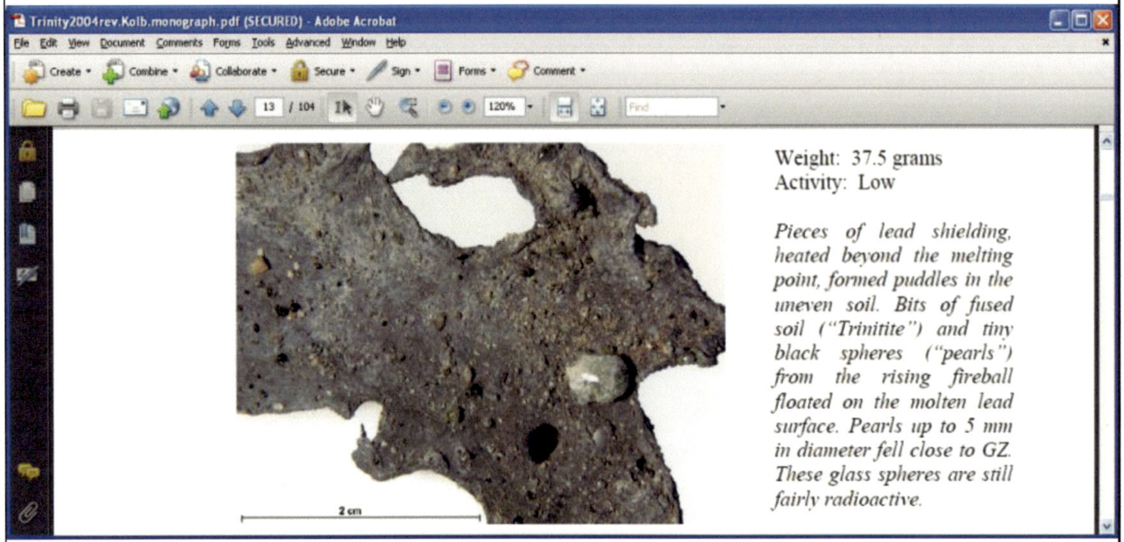

Item kindly donated by Harry Baldwin for the BSM display
Photo: courtesy of Bill Kolb

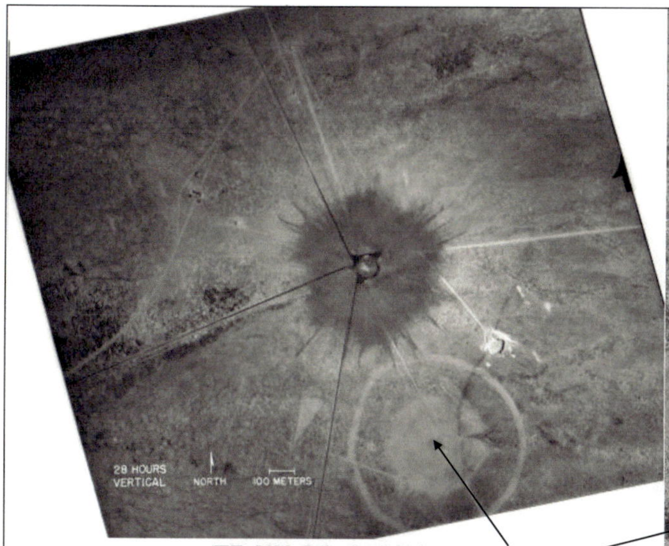

TR-NN-06_8x10.jpg

100 ton

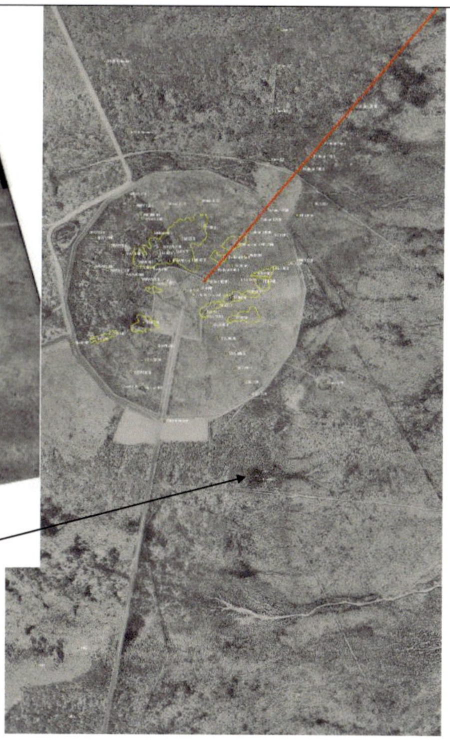

Scaled view from above

Red line is primary direction of fall-out

puddles of Trinitite

“colorized” by Omar J. (LANL)

http://fas.org/nuke/hew/usa/tests/trinity.html

Where did the Trinitite go?

Ralph Pray defeated the gate locks in 1951
Courtesy R. Pray / B. Kolb

Ralph liberated about 2000 lbs of Trinitite in five trips

Dealers in Santa Fe, Ojo Cliente and Corona peddled this Trinitite for many years

PILFERED by lots of people!

The remaining 500 pounds is now owned by Derik Bower in Hagerstown

Where did the Trinitite go?

In 1967, the "hottest" Trinitite was placed into 30 gallon ash cans, and transported to Los Alamos for burial !!!!!! **

** ***not*** in Area "B" on south side of DP road
Probably in Area "C", just south of TA-50

AEC photos courtesy of WSMR public affairs

Trinitite inside the shed located within the inner fence.

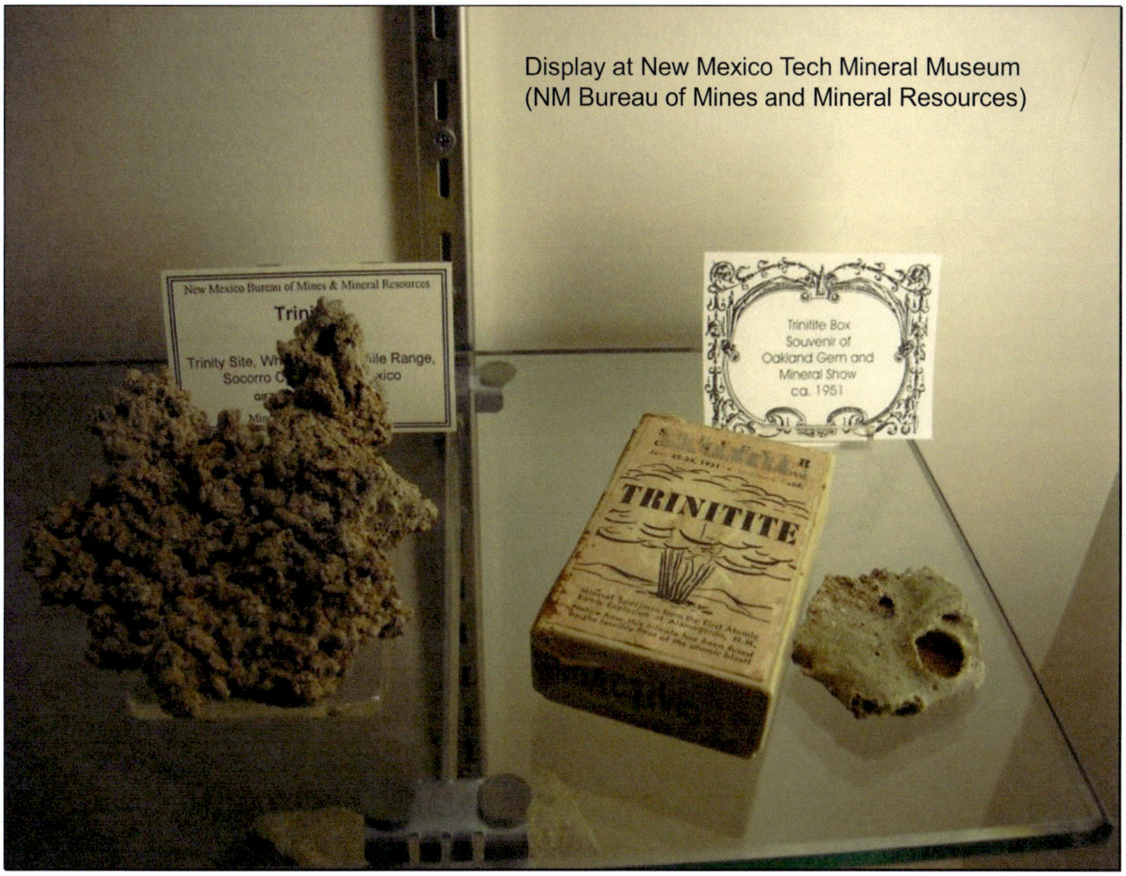

Harry Baldwin collection

Beaded bottom of the Green Monster (18 cm or 7" across)

Alternate forms of Trinitite

beaded "globs"

"taffy" ribbon

Collection and photos by Robb Hermes (with permission from WSMR)

"RED" TRINITITE

“Pancake” Trinitite Radioactivity

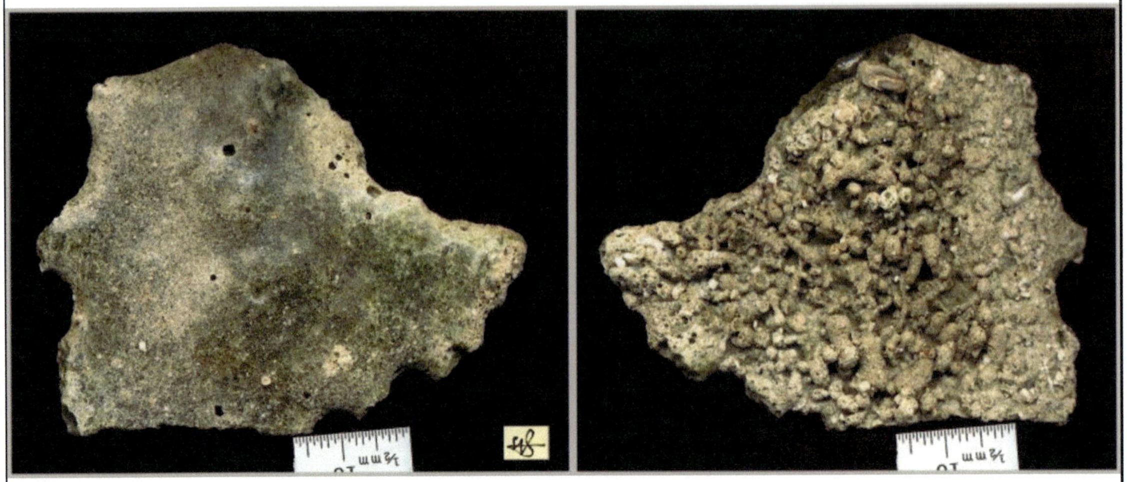

Mr. Bill’s Trinitite Pancake

* reading with CDV-700 (beta/gamma)

Top	1.5 mr/hr ~1000 cpm
Bottom	.3-.4 mr/hr 250-300 cpm
Background	.1 - .2 mr/hr 100-150 cpm

W. Strickfaden photos

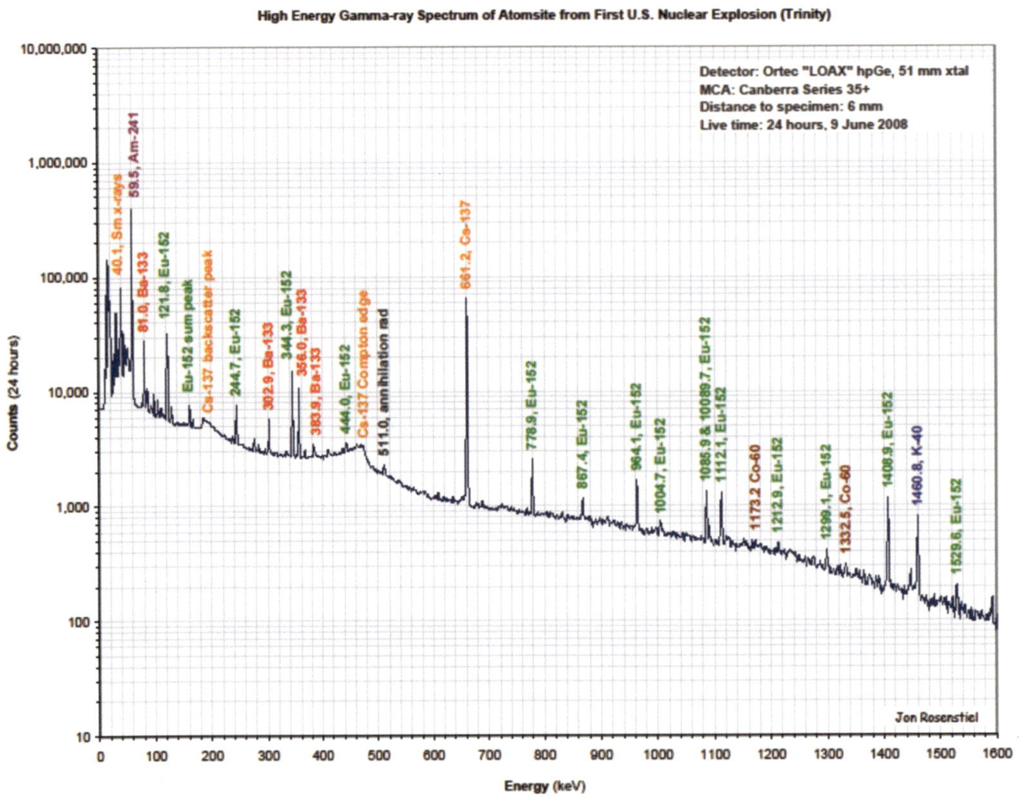

Small ant hill at Trinity Site - showing fragments of Trinitite
(nocturnal ants: *Myrmecocystus mexicanus) a.k.a. honey pot ants*

Bill Kolb photo
Separating ant rocks from Trinitite is tedious work

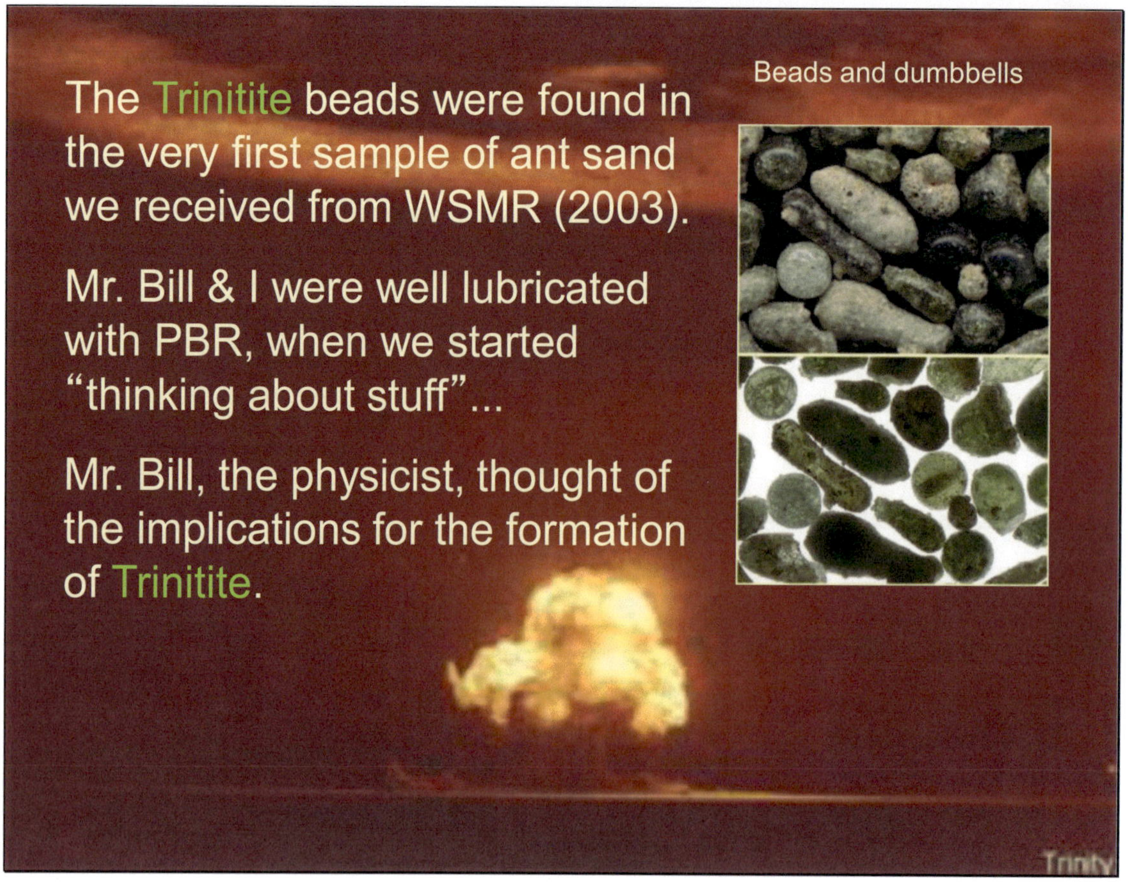
Beads and dumbbells
The Trinitite beads were found in the very first sample of ant sand we received from WSMR (2003).
Mr. Bill & I were well lubricated with PBR, when we started "thinking about stuff"...
Mr. Bill, the physicist, thought of the implications for the formation of Trinitite.
Trinity

"Thinking about stuff...the Trinitite particles found around ant-hills occur as beads and dumbbells and resemble tektites. These particles were apparently transported through the atmosphere as molten blobs.
Beads and dumbbells
Tektites
Trinitite dumbbell
Trinity

The Saga of the Ant Sand; more collecting
Photo by R. Hermes
Photo by O. Juveland

Plethora of Trinitite "beads" , ring is ~ 41 mm diameter

reflected light transmitted light

Photos by W. Strickfaden

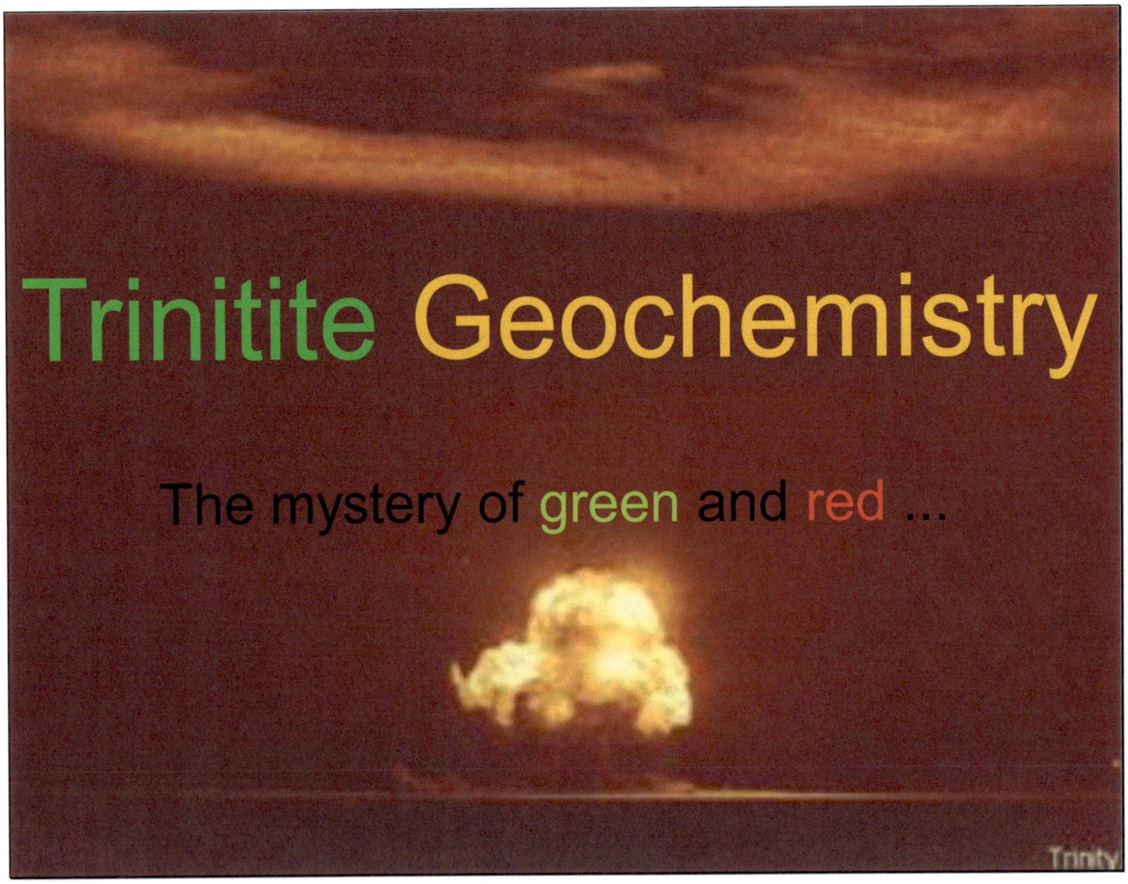

Trinitite: Red with Green coating

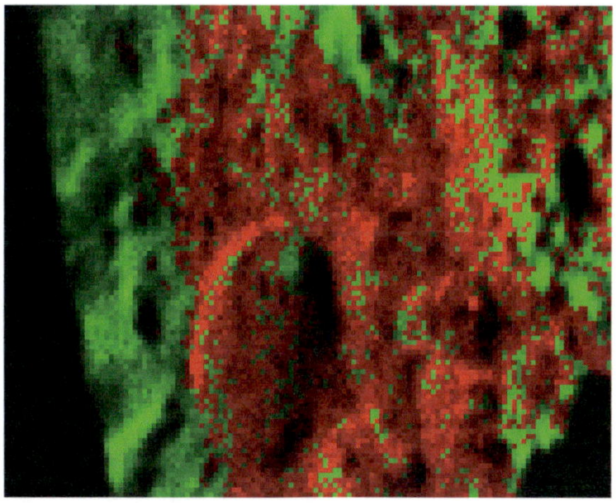

Fe/Cu overlay

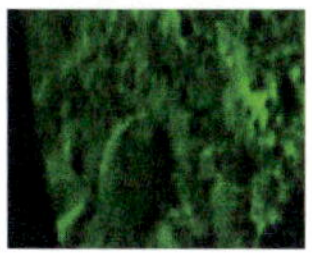

Fe map

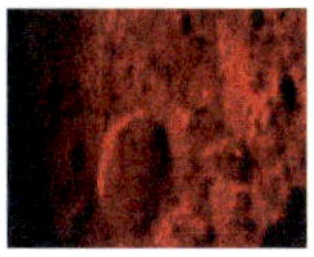

Cu map

Micro XRF courtesy of George Havrilla (LANL)

MXRF Spectra for TG14 BLOCK – Clear Trinitite bead

Element	Intensity	Wt%
AlK	259	14
SiK	1467	45
K K	1123	23
CaK	334	7
FeK	752	6

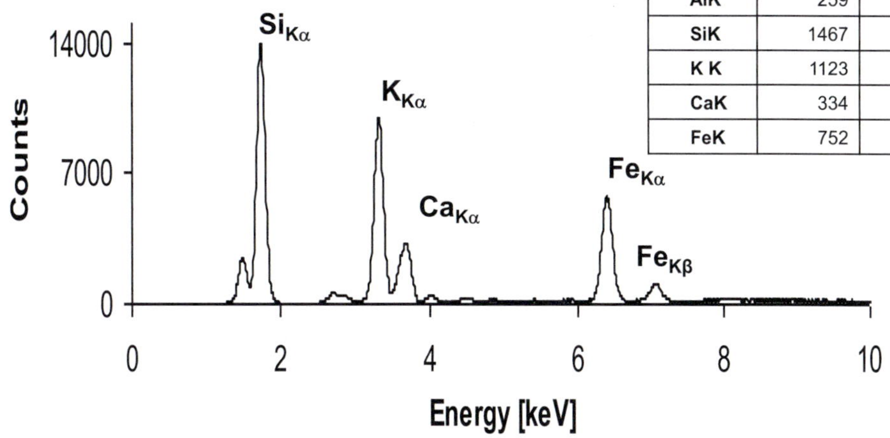

Micro XRF courtesy G. Havrilla and Velma Montoya (LANL)

TG14 BLOCK – Clear Trinitite bead Micro XRF Elemental Images

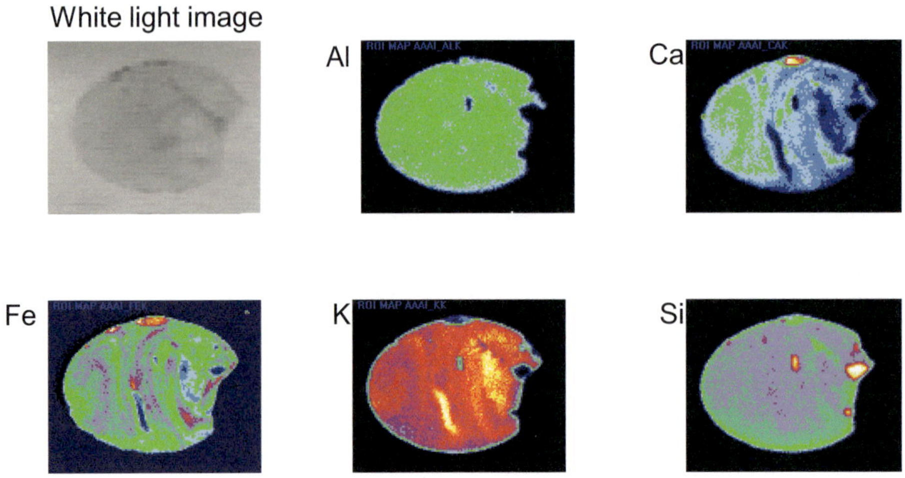

Note: K and its salts give a violet color flame

Scan distance (mm) 4 x 4
Step Size (μm) 29 x 39

Micro XRF courtesy G. Havrilla and Velma Montoya (LANL)

Heat of the blast melted the sand in place.
Trinity

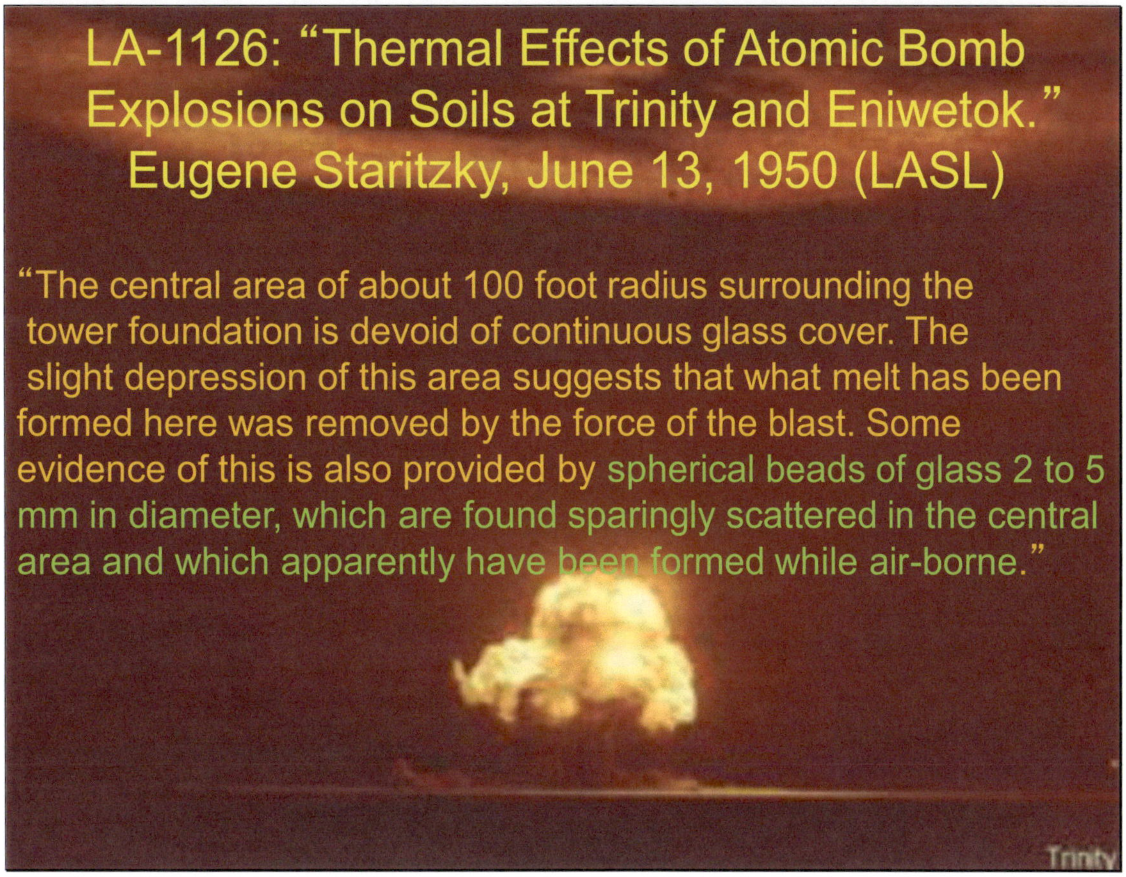
LA-1126: "Thermal Effects of Atomic Bomb Explosions on Soils at Trinity and Eniwetok."
Eugene Staritzky, June 13, 1950 (LASL)
"The central area of about 100 foot radius surrounding the tower foundation is devoid of continuous glass cover. The slight depression of this area suggests that what melt has been formed here was removed by the force of the blast. Some evidence of this is also provided by spherical beads of glass 2 to 5 mm in diameter, which are found sparingly scattered in the central area and which apparently have been formed while air-borne."
Trinity

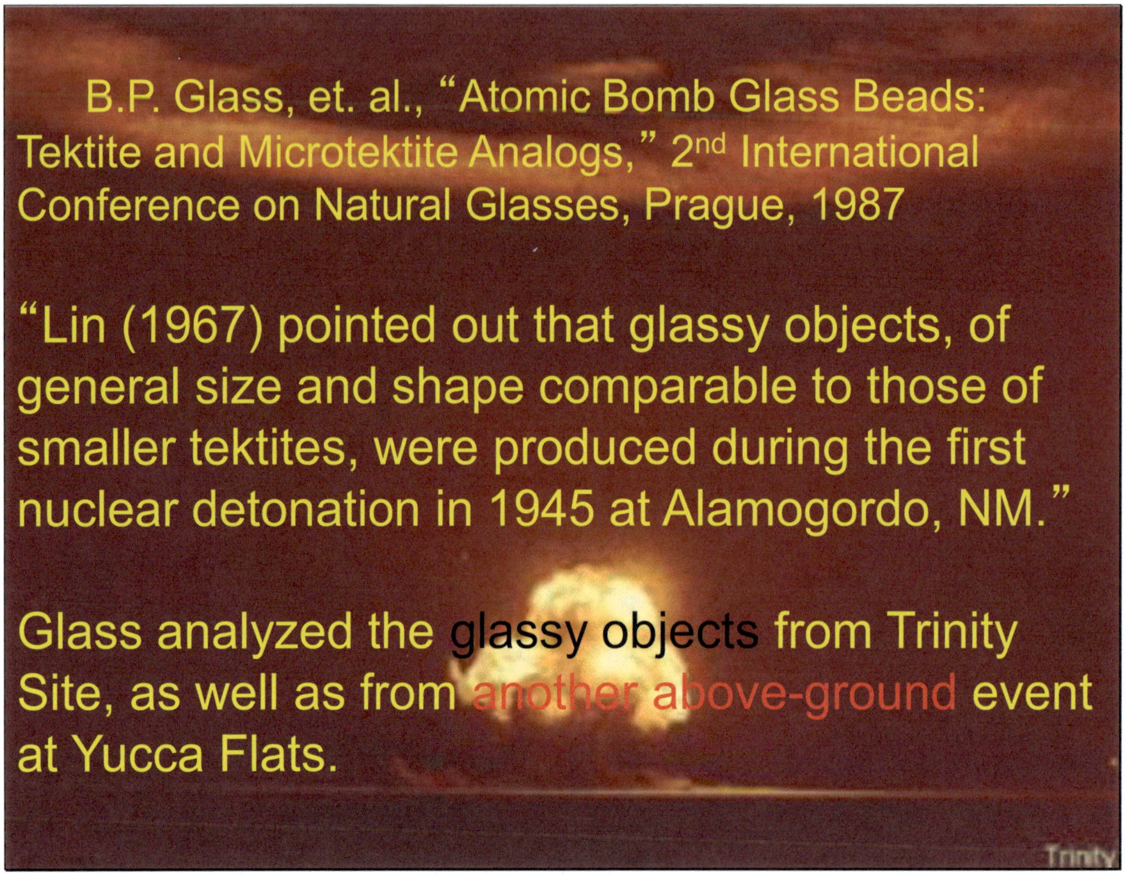

Sunday, April 23, 2006 THE NEW MEXICAN C-5

Scientists debunk theory of Trinity Site glass

By Sue Major Holmes
The Associated Press

ALBUQUERQUE — It was a rain of molten sand in the desert.

Trinitite — the smooth, jade-green glass formed by the world's first atomic-bomb test 60 years ago at Trinity Site — wasn't formed the way it has been explained for years, according to two scientists whose hobby is studying sand from anthills.

The first atomic detonation July 16, 1945, in the far reaches of what is now White Sands Missile Range, turned the sand in the blast area into a glassy, somewhat radioactive substance that became known as trinitite.

The story for decades was that the sand was heated instantaneously to nearly 5,000 degrees Fahrenheit, melting it.

But that's not what happened, say Los Alamos National Laboratory chemist Robert Hermes and Bill Strickfaden, a retired physicist for a Navy contractor and a student at the lab years ago.

According to their study of trinitite and little beads of trinitite found in anthills at Trinity Site, the sand was blasted into the air at superheated temperatures.

"Independently, the thing we came up with was a lot of the material that came from the crater got thrown up in the air so hot it melted in the air,"

Hermes said. "It rained down as these little droplets."

The ground itself was so hot the sand puddled together when the droplets hit, he said.

"The gist is the puddling came from the raining down of droplets ... rather than the heat just melting everything on the ground," Hermes said.

The idea is confirmed by the fact that trinitite is more radioactive on the top, glassy side than on the bottom, sandlike side. The blast threw a huge radioactive cloud into the air, but the ground where the material fell was comparatively not that radioactive, the pair said.

"We can essentially prove this because we found the beads themselves are a lot more radioactive than the trinitite on the ground," Strickfaden said.

Their work was published in the fall 2005 edition of *Nuclear Weapons Journal*, which actually was printed in February.

This new tidbit of nuclear history began with a tourist trip to Trinity Site, the test spot for the Los Alamos scientists who developed the bomb during World War II.

Hermes, who had lived in Los Alamos for 30 years, had never toured the site and decided to join one of the tours the missile range holds twice a year. While there, he noticed a lot of anthills and wondered if they contained trinitite.

Today, trinitite is considered a historic artifact, and people aren't allowed to pick it up. So after the tour, Hermes contacted Jim Eckles, a public-affairs officer for the range for nearly 30 years. Eckles obligingly sent material for the scientific study; later, a small group from Los Alamos was allowed to gather more.

According to the study, sand was blasted into the air at superheated temperatures when the world's first atomic bomb was detonated in 1945, and droplets of the smooth, jade-green glass rained down on the site.

Eckles believes it took so many decades to study trinitite "because nobody gave it a second thought. The common-sense talk you hear about the sand — it was hot, it melted and turned to glass — nobody had thought about it."

Hermes and Strickfaden said in their paper they wanted to use the properties of trinitite to back-calculate the yield of the bomb, the temperature and duration of its fireball and the spread of material ejected by the blast.

But their study also suggested trinitite was deposited by a rain of molten droplets and puddling from the heat. The paper in the *Nuclear Weapons Journal* — the closest edition to the 60th anniversary of the bomb — is a condensed version of their findings.

Their yield calculations are within those published for years and their temperature calculation "compares favorably" with that published by the Atomic Energy Commission, they wrote.

Strickfaden calculated the fireball's temperature at 8,430 degrees Kelvin — more than 14,000 degrees Fahrenheit.

The trinitite beads that the two found "of extreme interest" generally were 2 to 3 millimeter spheroids, Hermes said. Some are beautifully round; others are teardrops or oblong; some are shaped like tiny dumbbells.

"It's a fascinating side trip into physics," he said.

The pair did their research on the side, and it was not funded by the lab.

"It started with our first love, which was to investigate the mineralogy of ant sand," Hermes said.

Associated Press story (result of NWJ article):
April 23, 2006 - The New Mexican, and
all over the country!

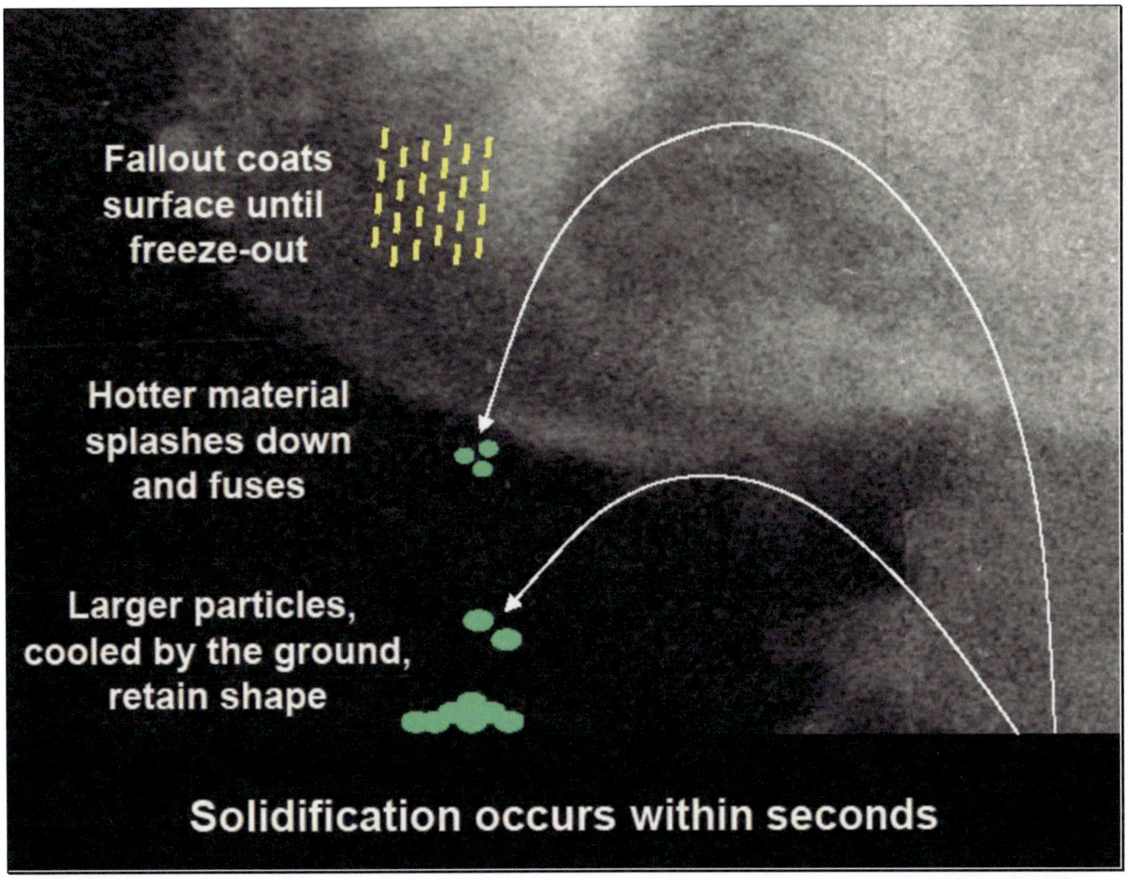

The New Model

Shockwave fractures rock under tower, compresses earth and scours crater

Ground springs back, ejecting material upward from crater

Ejecta thrown out at low angles rains down

Ejecta blown upward into fireball melts, mixes with fission products and splashes down

Fallout fuses on cooling surfaces

Evidence that Trinitite wasn't formed in-situ

- Asphalt surfaces near GZ were unmelted
- Intact asphalt was covered with Trinitite
- Heavy Trinitite deposits were found on wood and plant remains
- Rocks, pieces of metal, and a concrete pillar two feet high were covered with Trinitite
- Small pieces of Trinitite were found up to a mile away

TR-NN-06_8x10.jpg

Trinity

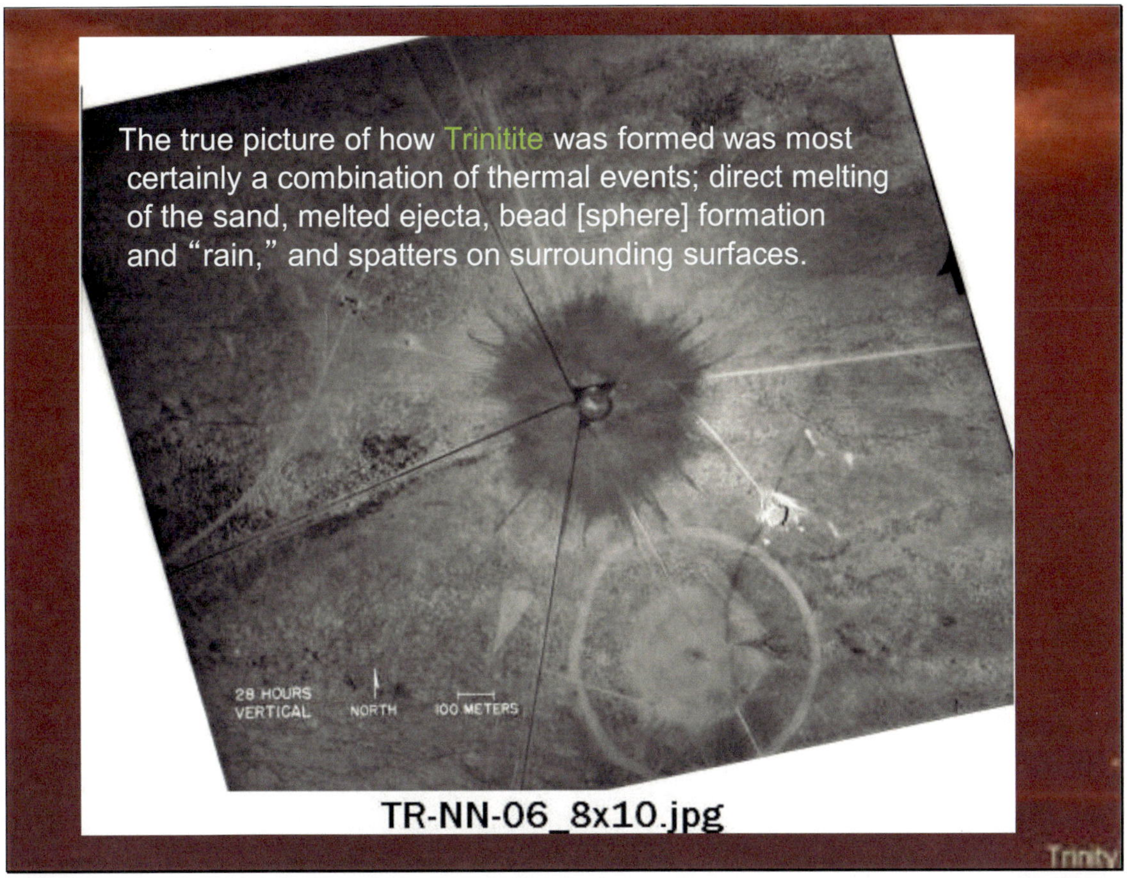
The true picture of how Trinitite was formed was most certainly a combination of thermal events; direct melting of the sand, melted ejecta, bead [sphere] formation and "rain," and spatters on surrounding surfaces.
28 HOURS VERTICAL
NORTH
100 METERS
TR-NN-06_8x10.jpg
Trinity

Other Interesting Stuff
Trinity

JOE-1

The nearly carbon copy of both the Trinity gadget and Fat Man!

The **RDS-1** (Russian: РДС-1), also known as *First Lightning* (Первая молния, *Pervaya molniya*), was the Soviet Union's first nuclear weapon test. In the west, it was code-named **Joe-1**, in reference to Joseph Stalin. It was test-exploded on 29 August 1949, at Semipalatinsk, Kazakh SSR, after a top-secret R&D project. Wikipedia

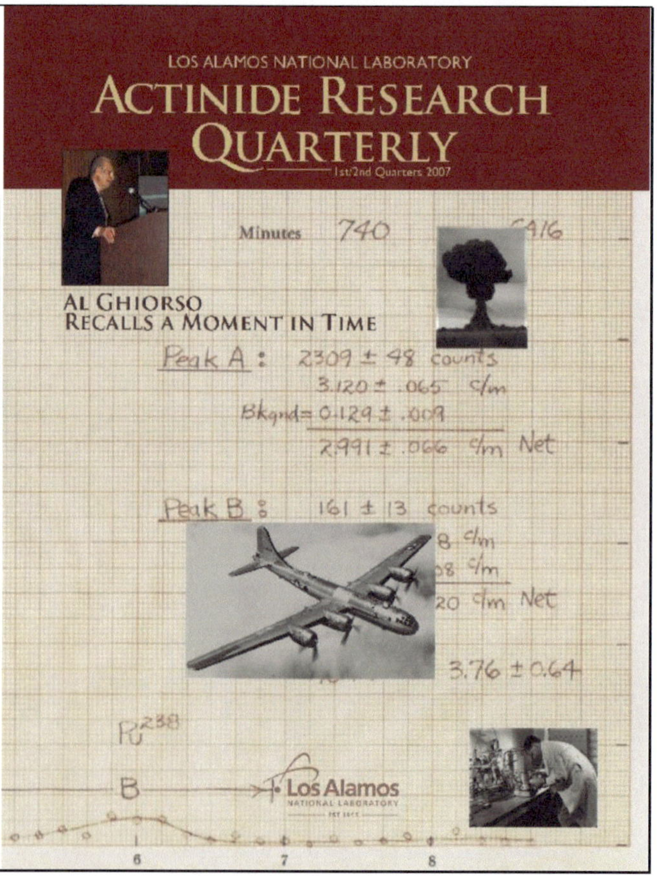

"Stalinite" (Joe-1 event) samples

Kharitonchik (plural: Kharitonchiki) is the name given to the pieces of molten rock left at ground zeroes after Soviet atmospheric nuclear tests at Semipalatinsk Test Site in Kazakhstan. This porous black material is named after one of the leading Russian nuclear weapons scientists, Yulii Borisovich Khariton. Wikipedia

shiny top

beaded bottom

Samples collected by Byron Ristvet, Ph.D., DTRA, ABQ;
with permission of the Kazakhstan National Nuclear Center
and the Russian Federation Atomic Energy Agency, Oct. 2006

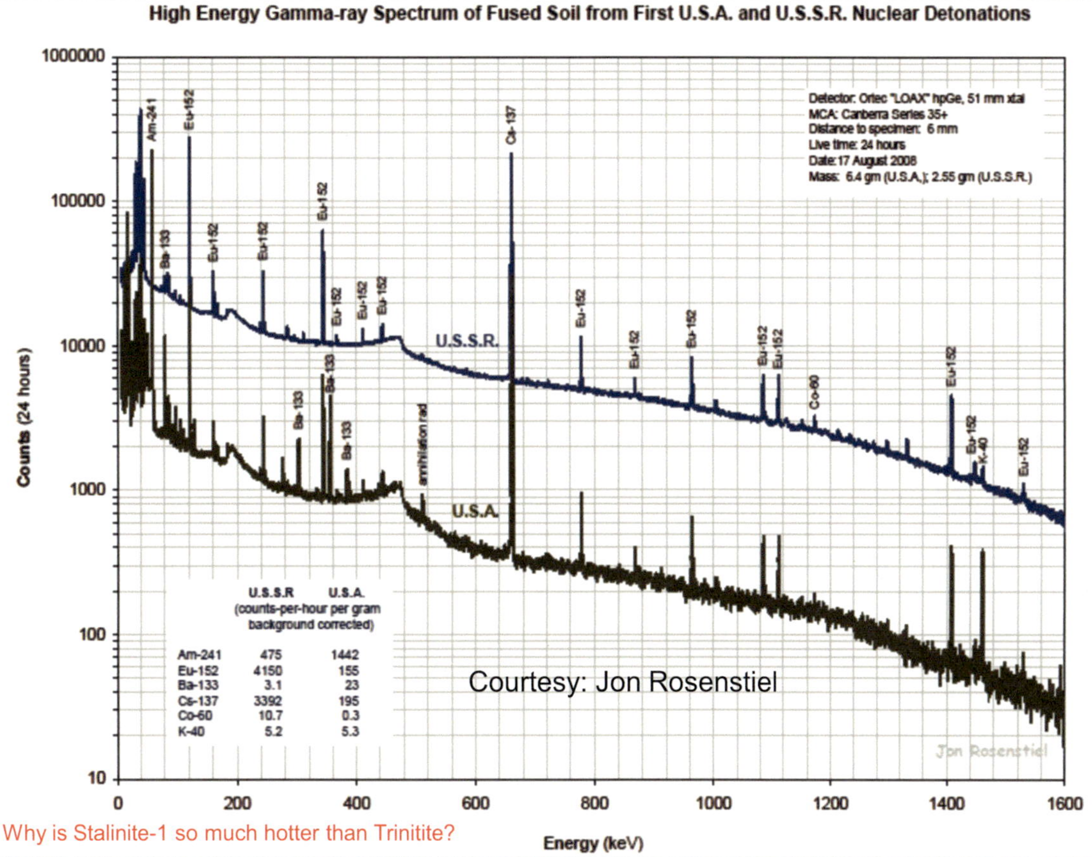

Why is Stalinite-1 so much hotter than Trinitite?

Joe 4 (Warhead name: RDS-6s (*Reaktivnyi Dvigatel Specialnyi*; Special Jet Engine)) was an American nickname for the first Soviet test of a thermonuclear weapon on August 12, 1953. It utilized a scheme in which fission and fusion fuel (lithium-6 deuteride) were "layered", a design known as the ***Sloika*** (Russian: Слойка, named after a type of layered puff pastry) model in the Soviet Union.

Joe 4 detonated with a force equivalent to 400 kilotons. (a combination of fission and fusion)

Source: Wikipedia

Joe-1/4 site Brian Rees Sept. 2011

Bunker at Joe-1/4 site Brian Rees Sept. 2011

Bunker near Joe-1/4 site Brian Rees – Sept. 2011

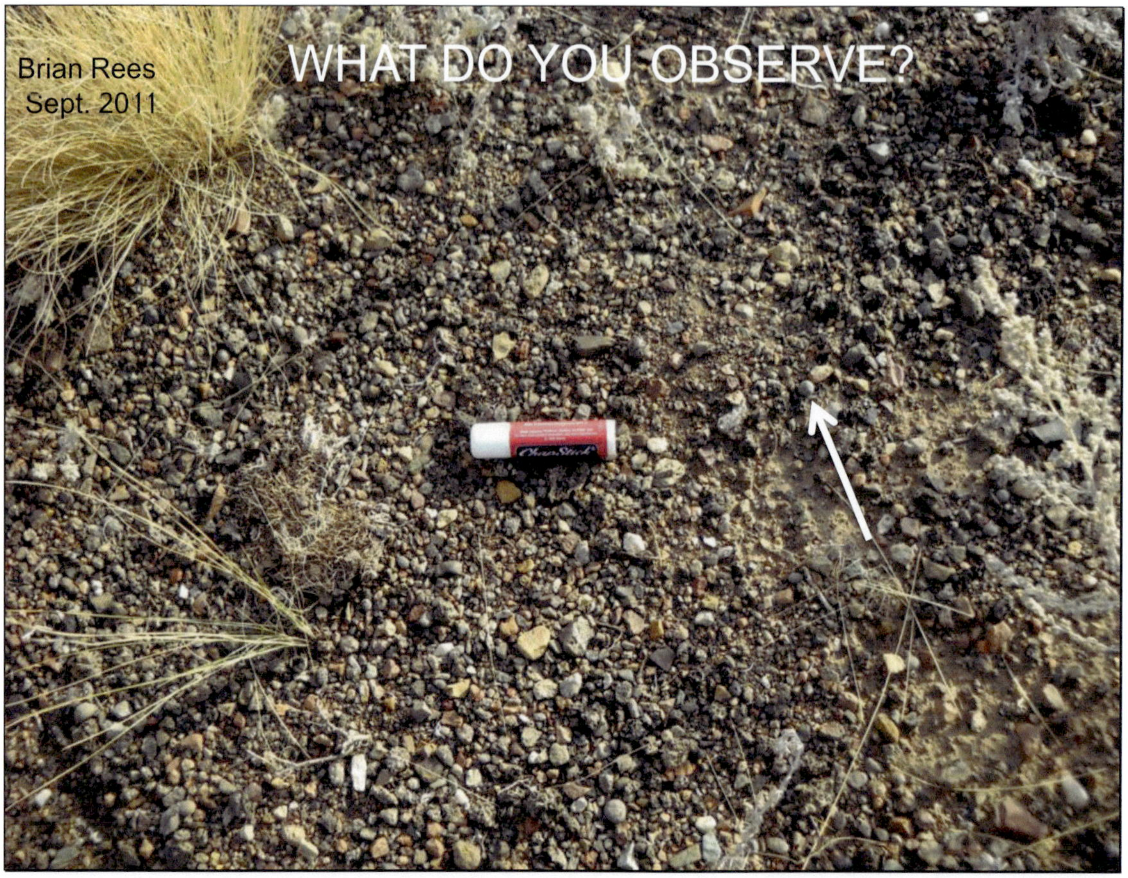
Brian Rees
Sept. 2011
WHAT DO YOU OBSERVE?

BRADBURY
SCIENCE
museum
What is this Laboratory scientist finding so intriguing about the sand at Trinity Site?
A New Look at Trinitite
Our new exhibit describes how ants, ant sand, and the curiosity of two scientists contributed to a new theory about the creation of trinitite.
Do you own a sample of trinitite? Bring it in to compare it with the different varieties of trinitite on display. Our new exhibit also includes examples of fused soil from natural events.
LOS ALAMOS NATIONAL LABORATORY
HOURS:
Tuesday–Saturday, 10–5
Sunday & Monday, 1–5
505-667-4444
www.lanl.gov/museum
10.01.2011 13:47

A New Look at Trinitite
09 29 2011 15:21

Trinitite Geochemistry
And the real nitty gritty for you geochemists!
Trinity

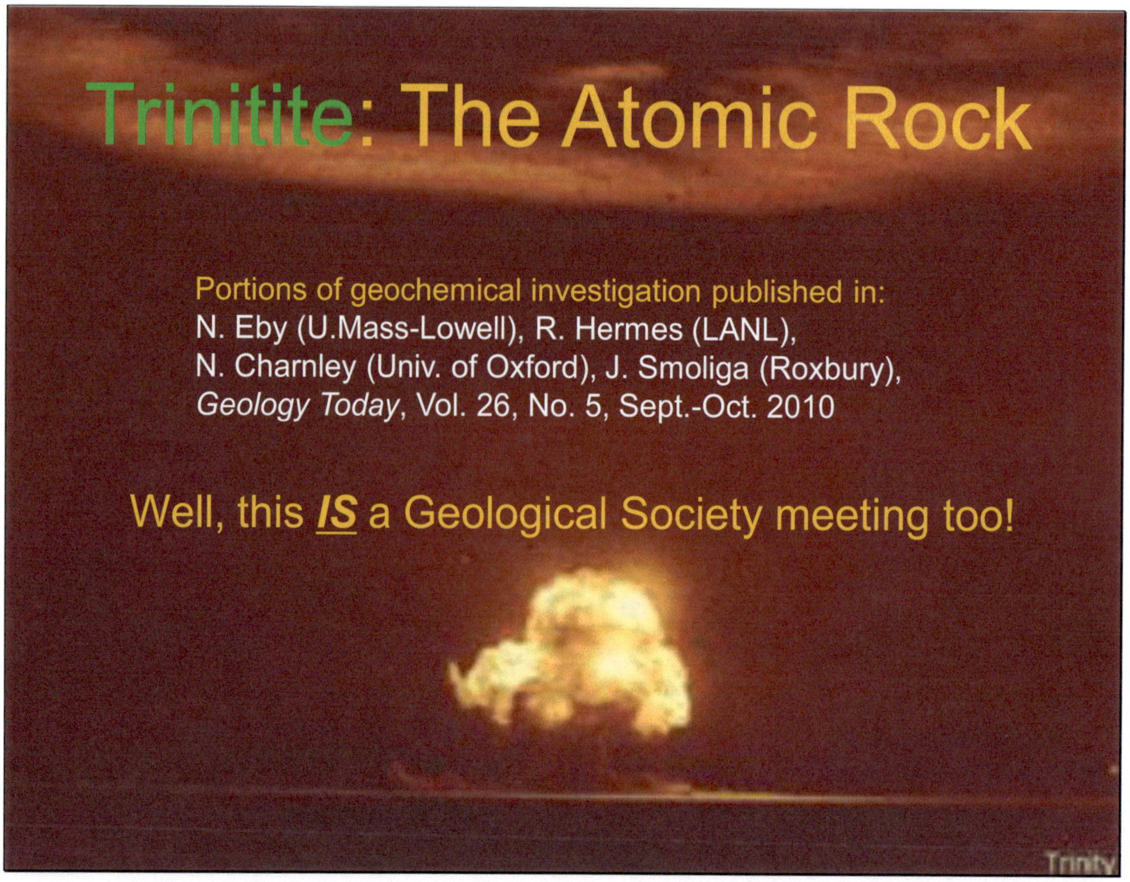
Trinitite: The Atomic Rock
Portions of geochemical investigation published in:
N. Eby (U.Mass-Lowell), R. Hermes (LANL),
N. Charnley (Univ. of Oxford), J. Smoliga (Roxbury),
Geology Today, Vol. 26, No. 5, Sept.-Oct. 2010
Well, this IS a Geological Society meeting too!
Trinity

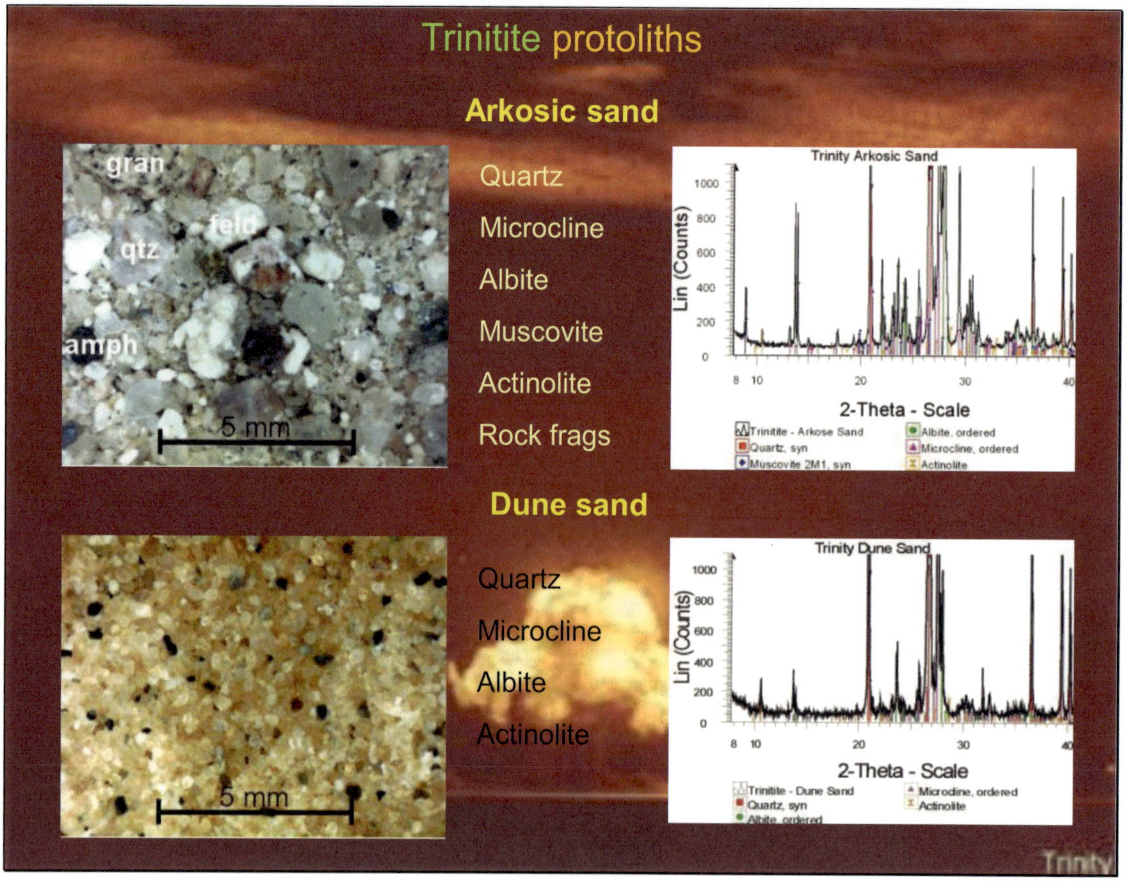
Trinitite protoliths
Arkosic sand
gran
felo
qtz
amph
5 mm
Quartz
Microcline
Albite
Muscovite
Actinolite
Rock frags
Trinity Arkosic Sand
Lin (Counts)
0
200
400
600
800
1000
8
10
20
30
40
2-Theta - Scale
Trinitite - Arkose Sand
Quartz, syn
Muscovite 2M1, syn
Albite, ordered
Microcline, ordered
Actinolite
Dune sand
5 mm
Quartz
Microcline
Albite
Actinolite
Trinity Dune Sand
Lin (Counts)
0
200
400
600
800
1000
8
10
20
30
40
2-Theta - Scale
Trinitite - Dune Sand
Quartz, syn
Albite, ordered
Microcline, ordered
Actinolite
Trinity

XRD patterns for Trinitite fragments and beads

The only crystalline phase found in Trinitite is alpha-quartz. All other phases were melted during the detonation.

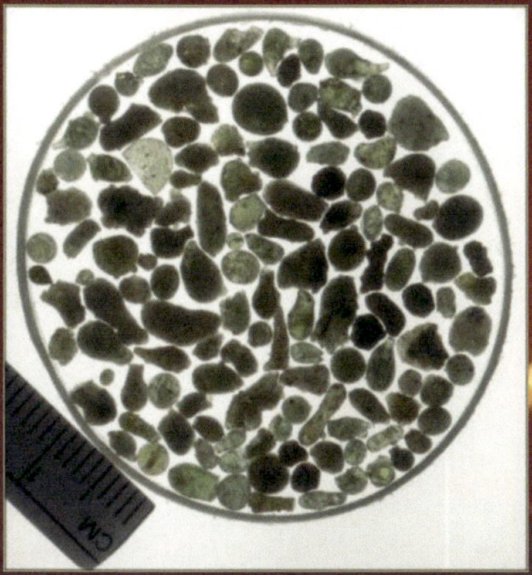

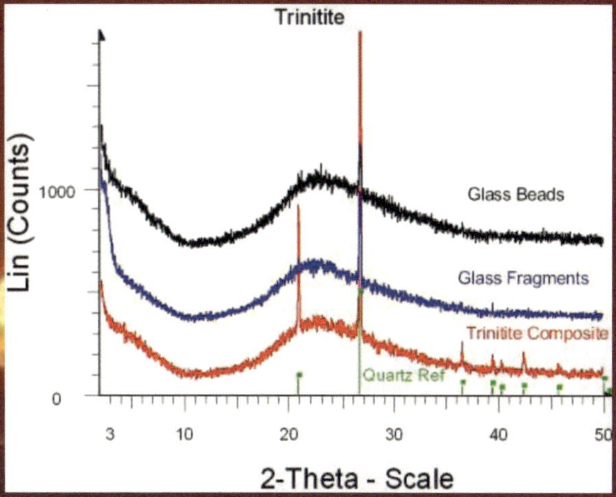

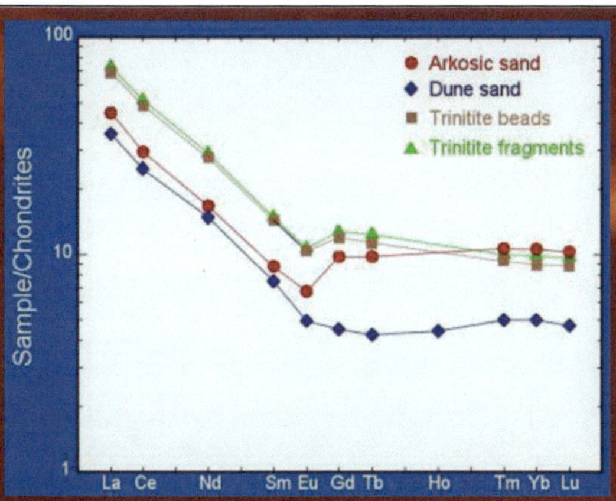

Chemical changes from source material (sand) to Trinitite

(Neutron Activation Analysis)

Trace metals, Sc, Co and Ni increase in the Trinitite.

Fission products, Ba, La, etc. increase in the Trinitite.

	Trinite Fragments	Trinititie Beads	Fine sand	Dune sand
Sc	4.48	4.9	1.9	2.16
Co	4.7	5.34	1.86	2.63
Ni	23	21	8.8	7.0
Sr	179	176	126	80
Ba	808	740	694	432
La	24.13	22.67	14.74	11.84
Gd	3.5	3.3	2.7	1.25
Yb	2.16	1.97	2.33	1.1
Hf	6.91	5.19	4.96	4.06
Ta	0.78	0.67	0.44	0.35
Th	9.15	8.59	6.99	2.01
U	2.7	2.62	1.8	0.75

Th and U increase in Trinitite probably from bomb. U was used in the tamper.

HREE* are similar in Trinitite and sand. Not added during the nuclear detonation.

*Heavy Rare Earth Elements

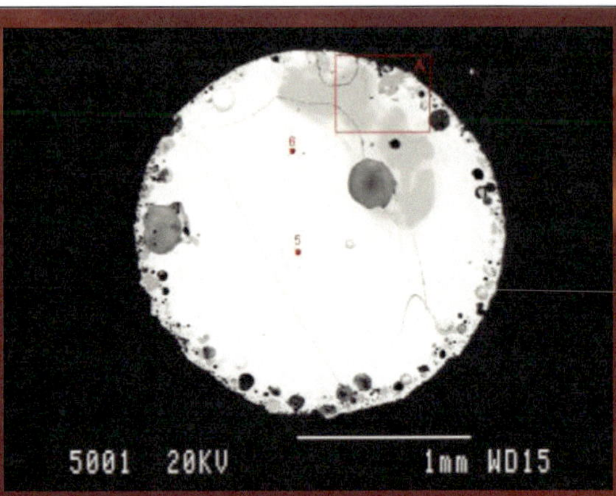

Trinitite Bead

Note quartz grains embedded along edge of bead. These were entrained during transport.

Darker areas within the bead are melted quartz grains (TB1). Also note the range in chemical composition shown for the glasses even in this small bead. TB4 is melted K-feldspar. High CaO in other glasses may be contributed by melted carbonate grains.

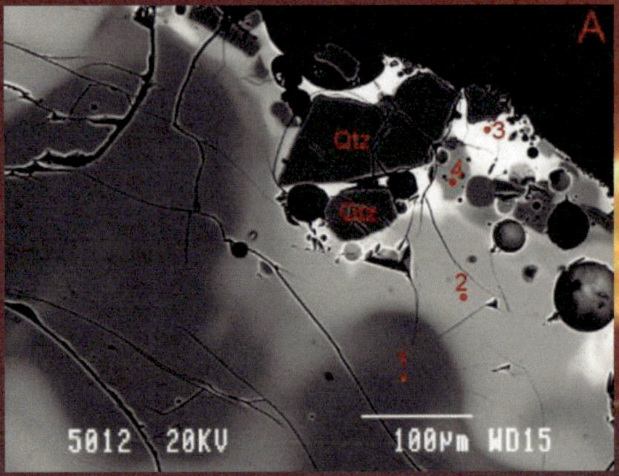

	TB1	TB2	TB3	TB4	TB5	TB6
SiO2	96.80	64.31	51.53	63.56	62.60	64.02
TiO2	0.10	0.31	0.26	0.08	0.37	0.27
Al2O3	0.35	14.57	11.49	18.46	15.19	14.67
FeO	0.17	2.18	3.20	0.13	2.39	2.37
MnO	nd	0.13	0.11	nd	nd	0.09
CaO	1.40	12.70	27.33	0.52	12.81	12.91
MgO	0.10	1.10	1.78	0.29	1.27	1.10
Na2O	0.10	0.95	0.73	1.83	0.74	0.74
K2O	0.53	1.44	2.66	13.47	1.43	1.44
Total	99.56	97.68	99.09	98.33	96.80	97.62

Trinity

Variations in Trinitite Chemistry – Glass Fragment

|---------------------- 0.6 mm -------------------|

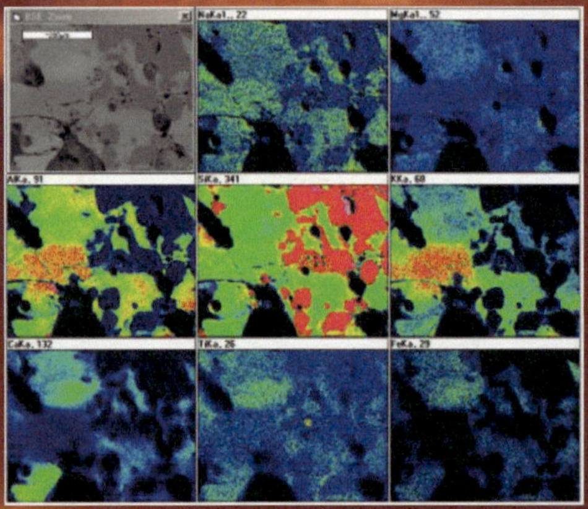

Sample	SiO2	TiO2	Al2O3	FeO	MnO	MgO	CaO	Na2O	K2O	Total
GL9 (light)	56.75	1.89	13.47	4.29	0.10	2.28	12.53	1.22	4.54	97.07
GL10 (dark)	63.46	0.27	18.82	nd	nd	0.21	0.96	2.08	11.70	97.50
GL11 (light)	60.21	0.51	15.40	1.86	nd	1.16	10.71	1.49	6.50	97.84
GL12 (med)	62.99	0.36	13.86	2.52	0.11	1.68	8.49	1.46	6.11	97.58
GL13 (med)	62.26	0.45	14.65	3.00	0.07	1.81	8.38	1.63	5.63	97.88
GL14 (med)	62.04	0.21	17.93	1.55	nd	1.05	6.89	1.97	6.16	97.80

Trinity

Trinitite Dumbbell

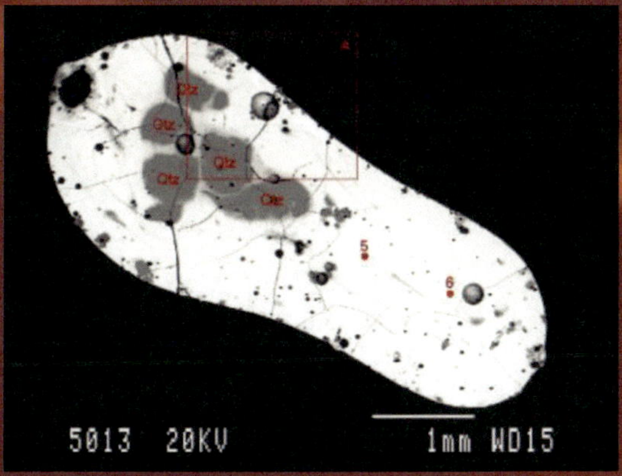

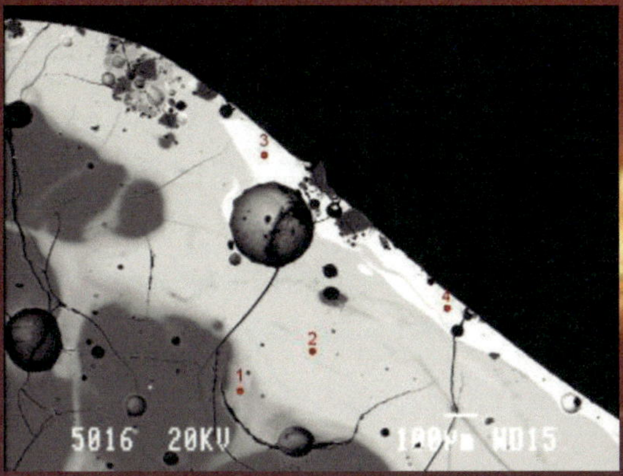

Glasses are MgO rich. TDB3 and TDB4, in terms of chemical composition, roughly correspond to Mg-Gedrite. Some of the other compositions suggest dilution by melted quartz grains plus feldspar. TDB1, which has the highest SiO2, is immediately adjacent to a partly melted quartz grain which supplied the SiO2.

	TDB1	TDB2	TDB3	TDB4	TDB5	TDB6
SiO2	70.53	64.41	47.52	47.15	64.25	66.32
TiO2	0.19	0.39	0.33	0.43	0.35	0.22
Al2O3	11.38	12.27	16.05	16.49	11.95	11.68
FeO	1.50	2.22	2.33	2.34	2.08	2.03
MnO	0.02	nd	nd	0.05	nd	0.06
MgO	10.66	15.50	31.17	30.88	15.44	13.68
CaO	1.01	1.21	1.56	1.55	1.04	1.10
Na2O	2.26	1.84	0.45	0.54	1.77	1.85
K2O	3.15	2.66	0.63	0.58	2.77	3.07
Total	100.68	100.52	100.05	100.02	99.66	100.00

Trinity

Geochemistry Conclusions

- Trinitite occurs in a variety of forms – (1) ~ 2 cm thick pancake trinitite with a glassy top and fused mineral grains constituting the bottom layer, (2) green vesiculated glass fragments, (3) red Trinitite (red color due to copper in the glass), and (4) glass beads and dumbbells.
- Residual radioactivity is still detectable in the glass. The source of the activity is Pu and U from the bomb, fission fragments (Cs, Ba, LREE), and neutron activation products (Co, Eu).
- The protolith for the Trinitite glasses is an arkosic sand that consists of quartz, microcline, albite, muscovite, actinolite, and carbonate. The only crystalline phase found in the glasses is alpha-quartz.
- Copper is found in the red trinitite glass and the glass also contains metallic chondrules consisting variously of iron, copper, and lead.
- At the 10s of micron levels the Trinitite glasses are very heterogeneous. They usually contain quartz grains and recognizable domains of melted quartz. Glass compositions range from almost pure silica to K-rich, Na-rich, and Ca-rich compositions. The Ca-rich glasses represent the addition of Ca from calcite grains and fossil fragments. None of the glass compositions represent equilibrium melts.

Trinity

Our Trinitite publications that led to further work!

Calculations (2005) | Geochemistry (2010)

Volcanic origin?
Trinity

Pu' u Pua' i, Kilauea, Hawaii 1959
sphere
plate
0 INCHES 1 2
0 CM 1 2 3 4 5 6 7
Bill Kolb collection
Pele' s tears

Volcanic activity can produce shiny surfaces with frothy interior

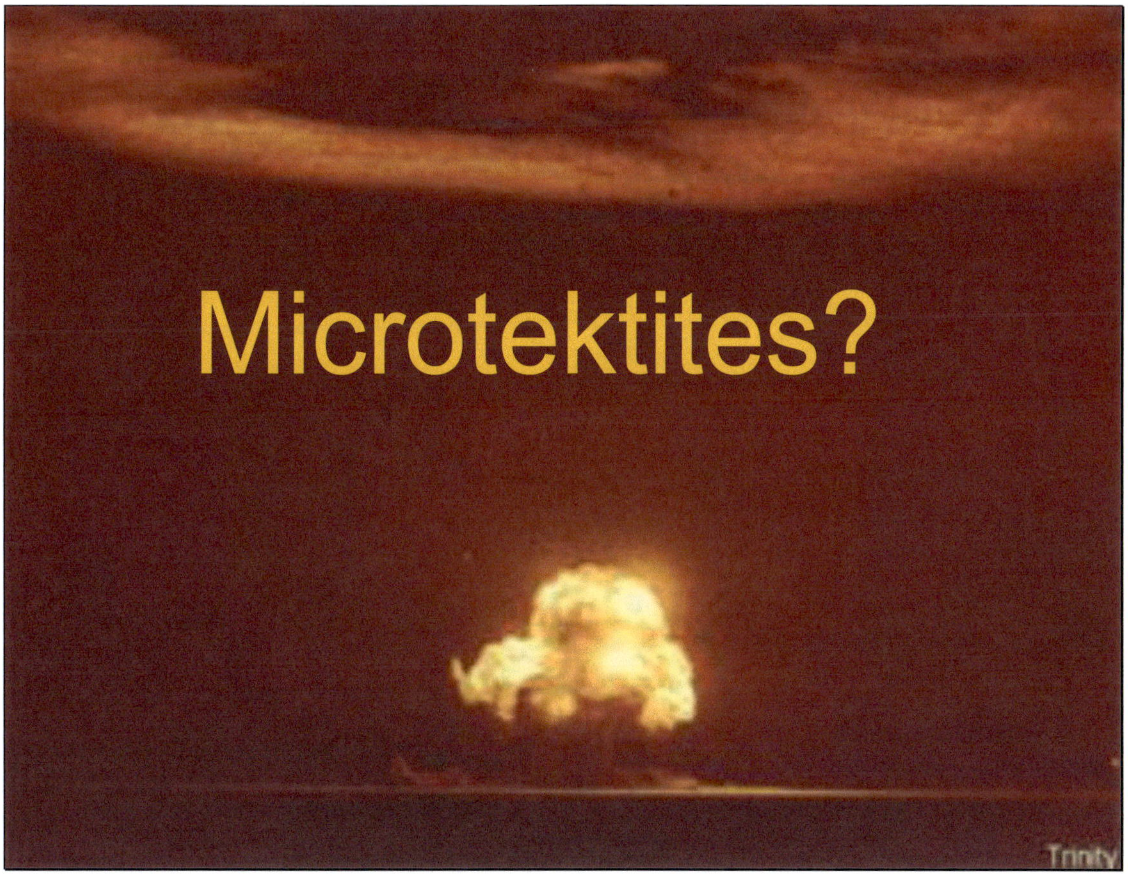

Meteor Crater, AZ (a.k.a. Barringer Impact Crater or Diablo Canyon Crater)

Impact spherules found near rim of Meteor Crater; scanned thin section on right.

Proposal for permission to collect ant sand from the area was denied ....

http://www.lpi.usra.edu/meetings/lpsc2005/pdf/1907.pdf

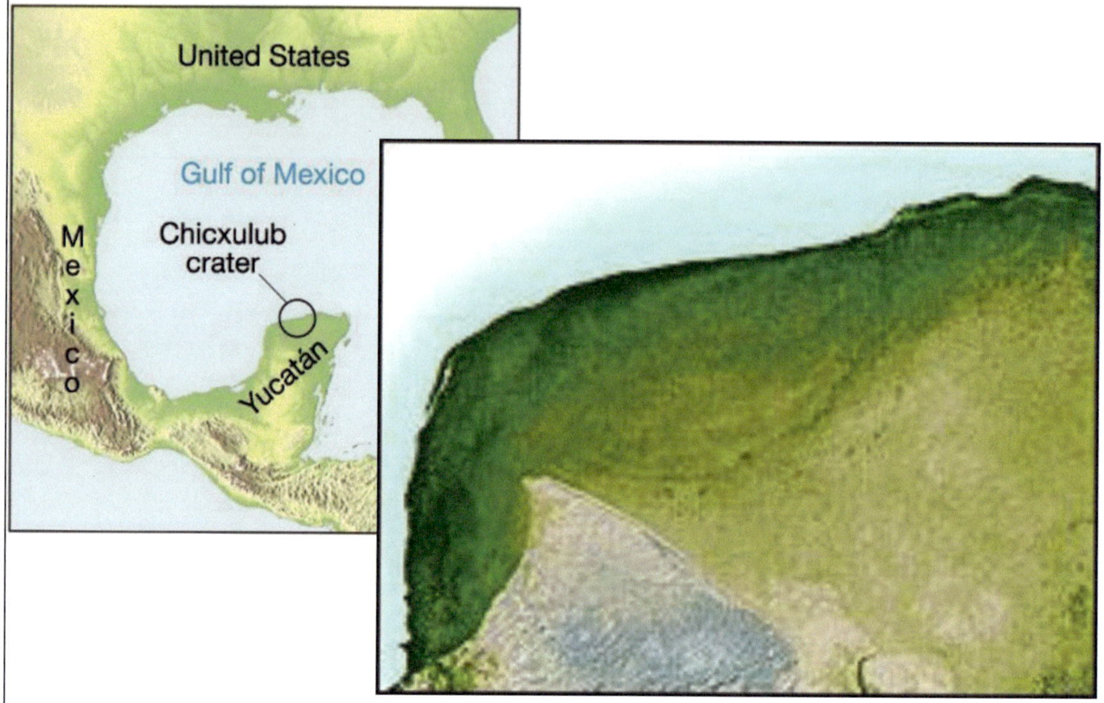

Chicxulub Event, ~ 65 million years ago

"Cretaceous-Tertiary (K-T) boundary"

Chicxulub microtektites collected by
Dr. David Kring, et.al., CSDP

Chicxulub Scientific Drilling Project
Separated from core obtained in GoM

Separated from slip fault exposure in Haiti (same fault that caused earthquake in January 2010)

Besides erosion, what other natural process happens here?

View of Trinity Site area from Atom Site (top of Oscuro Mountains)

"Atom Site" ant hill with "beads" (non-radioactive!)

"Atom Site" is a radar site on top of the Oscuro Mountains and in the fall-out path of the Trinity explosion.

Discovery and photo by Fred Yarger, MIT Lincoln Lab @ WSMR

Small ant hill at Atom Site, top of Oscuro Mountains

Exogenic Fulgurite?*
* Ejecta from a lightning strike!
Trinity

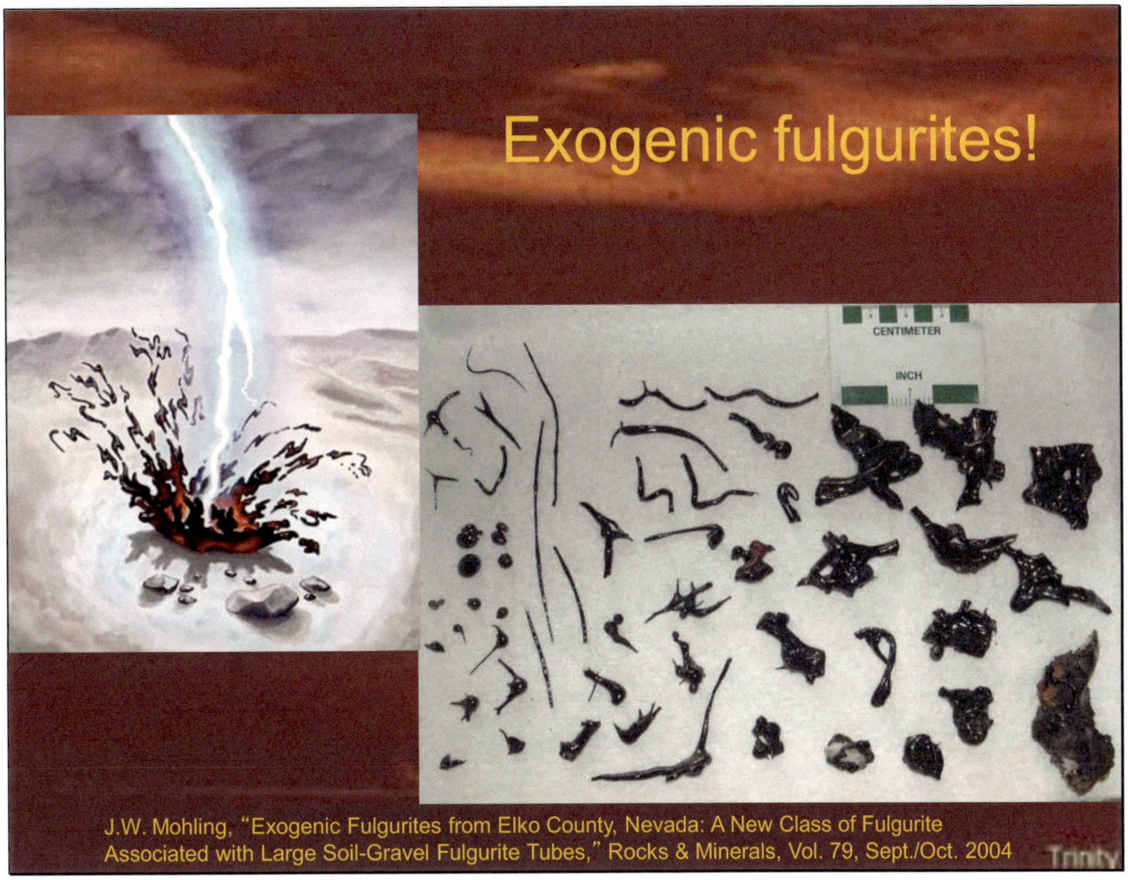
Exogenic fulgurites!
CENTIMETER
INCH
J.W. Mohling, "Exogenic Fulgurites from Elko County, Nevada: A New Class of Fulgurite Associated with Large Soil-Gravel Fulgurite Tubes," Rocks & Minerals, Vol. 79, Sept./Oct. 2004
Trinity

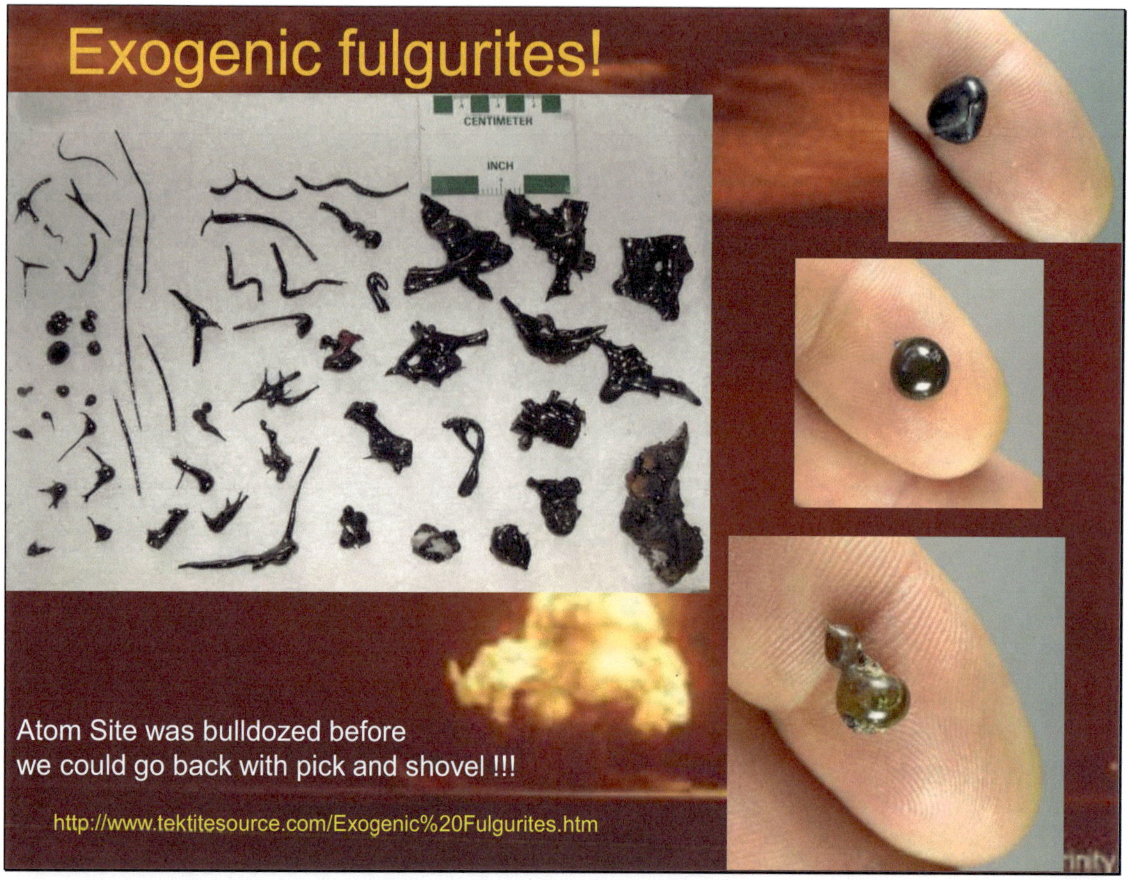
Exogenic fulgurites!
CENTIMETER
INCH
Atom Site was bulldozed before
we could go back with pick and shovel !!!
http://www.tektitesource.com/Exogenic%20Fulgurites.htm

Younger-Dryas
Extinction Event
Hypothesis
Trinity

Very high-temperature impact melt products as evidence for cosmic airbursts and impacts 12,900 years ago

Ted E. Bunch[a,1], Robert E. Hermes[b], Andrew M.T. Moore[c], Douglas J. Kennett[d], James C. Weaver[e], James H. Wittke[a], Paul S. DeCarli[f], James L. Bischoff[g], Gordon C. Hillman[h], George A. Howard[i], David R. Kimbel[j], Gunther Kletetschka[k,l], Carl P. Lipo[m], Sachiko Sakai[m], Zsolt Revay[n], Allen West[o], Richard B. Firestone[p], and James P. Kennett[q]

[a]Geology Program, School of Earth Science and Environmental Sustainability, Northern Arizona University, Flagstaff, AZ 86011; [b]Los Alamos National Laboratory (retired), Los Alamos, NM 87545; [c]College of Liberal Arts, Rochester Institute of Technology, Rochester, NY 14623; [d]Department of Anthropology, Pennsylvania State University, University Park, PA 16802; [e]Wyss Institute for Biologically Inspired Engineering, Harvard University, Cambridge, MA 02138; [f]SRI International, Menlo Park, CA 94025; [g]US Geological Survey, Menlo Park, CA 94025; [h]Institute of Archaeology, University College London, London, United Kingdom; [i]Restoration Systems, LLC, Raleigh, NC 27604; [j]Kimstar Research, Fayetteville, NC 28312; [k]Faculty of Science, Charles University in Prague, and [l]Institute of Geology, Czech Academy of Science of the Czech Republic, v.v.i., Prague, Czech Republic; [m]Institute for Integrated Research in Materials, Environments, and Society (IIRMES), California State University, Long Beach, CA 90840; [n]Forschungsneutronenquelle Heinz Maier-Leibnitz (FRM II), Technische Universität München, Munich, Germany; [o]GeoScience Consulting, Dewey, AZ 86327; [p]Lawrence Berkeley National Laboratory, Berkeley, CA 94720; and [q]Department of Earth Science and Marine Science Instititute, University of California, Santa Barbara, CA 93106

Bunch TE, et al. (2012), Proc. Natl. Acad. Sci. USA 109(28): E1903-E1912

It has been proposed that fragments of an asteroid or comet impacted Earth, deposited silica-and iron-rich microspherules and other proxies across several continents, and triggered the Younger Dryas cooling episode 12,900 years ago. Although many independent groups have confirmed the impact evidence, the hypothesis remains controversial because some groups have failed to do so. We examined sediment sequences from 18 dated Younger Dryas boundary (YDB) sites across three continents (North America, Europe, and Asia), spanning 12,000 km around nearly one-third of the planet. All sites display abundant microspherules in the YDB with none or few above and below. In addition, three sites (Abu Hureyra, Syria; Melrose, Pennsylvania; and Blackville, South Carolina) display vesicular, high-temperature, siliceous scoria-like objects, or SLOs, that match the spherules geochemically. We compared YDB objects with melt products from a known cosmic impact (Meteor Crater, Arizona) and from the 1945 Trinity nuclear airburst in Socorro, New Mexico, and found that all of these high-energy events produced material that is geochemically and morphologically comparable, including: (*i*) high-temperature, rapidly quenched microspherules and SLOs; (*ii*) corundum, mullite, and suessite (Fe_3Si), a rare meteoritic mineral that forms under high temperatures; (*iii*) melted SiO_2 glass, or lechatelierite, with flow textures (or schlieren) that form at >2,200 °C; and (*iv*) particles with features indicative of high-energy interparticle collisions. These results are inconsistent with anthropogenic, volcanic, authigenic, and cosmic materials, yet consistent with cosmic ejecta, supporting the hypothesis of extraterrestrial airbursts/impacts 12,900 years ago. The wide geographic distribution of SLOs is consistent with multiple impactors.

Fig. 1. Light photomicrographs of YDB objects. (*Upper*) SLOs and (*Lower*) magnetic spherules. **A** = Abu Hureyra, **B** = Blackville, **M** = Melrose.

Bunch TE, et al. (2012), Proc. Natl. Acad. Sci. USA 109(28): E1903-E1912

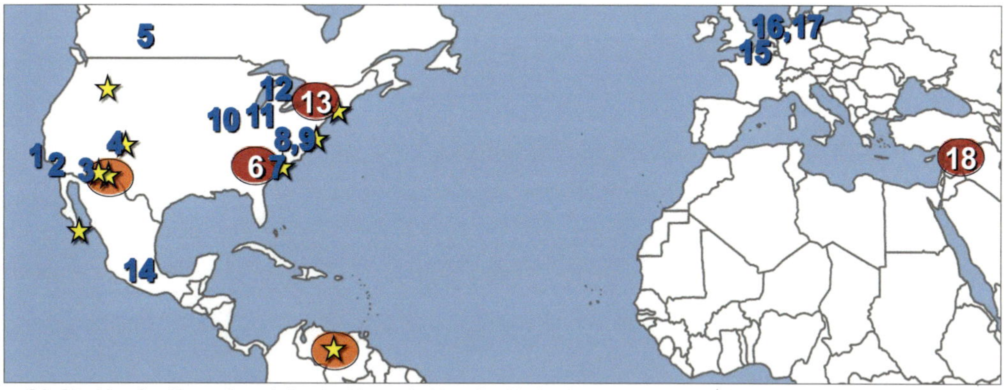

Figure S1. Site Map for 18 numbered sites in this study (described below), spanning about 12,000 km of the Northern Hemisphere. Currently, there is no known limit to the YDB impact field.

SLOs and spherules were observed at the three numbered sites in red circles: (#6) Blackville, South Carolina; (#13) Melrose, Pennsylvania; and (#18) Abu Hureyra, Syria.

Impact-related spherules without SLOs were observed at the other sites numbered in blue as follows: (#1) Arlington Canyon, California; (#2) Talega, California; (#3) Murray Springs, Arizona; (#4) Blackwater Draw, New Mexico; (#5) Chobot, AB, Canada; (#7) Topper, South Carolina; (#8) Barber Creek, North Carolina; (#9) Kimbel Bay, North Carolina; (#10) Big Eddy, Missouri; (#11) Sheriden Cave, Ohio; (#12) Gainey, Michigan; (#14) Cuitzeo, Mexico; (#15) Lommel, Belgium; (#16) Ommen, Netherlands; and (#17) Lingen, Germany.

Independent research. Also shown are 9 sites (yellow stars) investigated by 7 independent groups. SLOs were found at two sites, marked with a star in orange circle: Venezuela (Mahaney *et al.*, 2011) and Arizona (Fayek *et al.*, 2008, 2011). Impact-related YDB microspherules without SLOs were reported in the YDB layer in Montana (Baker *et al.*, 2008); Arizona (Haynes *et al.*, 2010); Mexico (Scruggs *et al.*, 2010); New Mexico, Maryland, South Carolina (LeCompte *et al.*, 2010); and Pennsylvania (Wu *et al.*, 2011).

Bunch TE, et al. (2012), Proc. Natl. Acad. Sci. USA 109(28): E1903-E1912

Figure S5. Melrose, PA. **A)** Area map showing location of the site (red star) about 1 km SW of the town of Melrose in northeastern PA (lat/long: 41.925350°N, 75.510066°W). The map shows the extent of glacial advance of the Laurentide Ice Sheet at approximately 25 ka (Fullerton *et al.* 2003) and began to retreat rapidly after 18 ka (Colgan *et al.*, 2003). **B)** Photo of exploratory trench showing lighter-colored glacial till (diamicton; below trowel), overlain by darker humic colluvium (photo credit: Malcolm LeCompte). The YDB is located in the colluvium in an 8-cm-thick layer from 15 to 23 cmbs, centered at 19 cmbs. **C)** Photomicrographs of subrounded SLOs ranging from about 800 to 5000 µm recovered from the YDB layer. These objects display evidence of melting at >2000°C, including lechatelierite (main paper, **Fig. 5C**), schlieren (**Fig. 16B**), impact pitting (**Fig. 6A**), and melt drapings (**Fig. 16D**) that conclusively rule out an origin by anthropogenesis, volcanism, or authigenesis. **D)** Linear interpolation is used to develop an age-depth model (red dotted line) based on one OSL date at 28 cmbs and the inferred age of surface sediments (**SI Table 2**). The vertical gray bar represents 12.9 ± 0.1 ka, and sediment sample locations are represented by purple diamonds. The green bar shows the depth of the peaks in SLOs and spherules in the 8-cm sample centered at 19 cmbs. Based on linear interpolation, the YDB layer is within the 12.9-ka age range.

Bunch TE, et al. (2012), Proc. Natl. Acad. Sci. USA 109(28): E1903-E1912

Figure S6. Light photomicrographs of magnetic and glassy spherules from Melrose, PA. Shapes include spherules, ovals, teardrops, and dumbbells. Colors include clear, gray, red, brown, and black. Note spherule **B** contains a large bubble. Both dumbbells (**D** and **H**) indicate fusion of molten or semi-plastic spherules. Note that dumbbell **H** consists of two dissimilar accretionary spherules, one clear (Si-rich) and the other opaque (Fe-rich).

Evidence for deposition of 10 million tonnes of impact spherules across four continents 12,800 y ago

James H. Wittke[a], James C. Weaver[b], Ted E. Bunch[a,1], James P. Kennett[c], Douglas J. Kennett[d], Andrew M. T. Moore[e], Gordon C. Hillman[f], Kenneth B. Tankersley[g], Albert C. Goodyear[h], Christopher R. Moore[i], I. Randolph Daniel, Jr.[j], Jack H. Ray[k], Neal H. Lopinot[k], David Ferraro[l], Isabel Israde-Alcántara[m], James L. Bischoff[n], Paul S. DeCarli[o], Robert E. Hermes[p,2], Johan B. Kloosterman[q,2], Zsolt Revay[r], George A. Howard[s], David R. Kimbel[t], Gunther Kletetschka[u], Ladislav Nabelek[u,v], Carl P. Lipo[w], Sachiko Sakai[w], Allen West[x], and Richard B. Firestone[y]

[a]Geology Program, School of Earth Science and Environmental Sustainability, Northern Arizona University, Flagstaff, AZ 86011; [b]Wyss Institute for Biologically Inspired Engineering, Harvard University, Cambridge, MA 02138; [c]Department of Earth Science and Marine Science Institute, University of California, Santa Barbara, CA 93106; [d]Department of Anthropology, Pennsylvania State University, University Park, PA 16802; [e]College of Liberal Arts, Rochester Institute of Technology, Rochester, NY 14623; [f]Institute of Archaeology, University College London, London WC1H0PY, United Kingdom; [g]Departments of Anthropology and Geology, University of Cincinnati, Cincinnati, OH 45221; [h]South Carolina Institute of Archaeology and Anthropology, University of South Carolina, Columbia, SC 29208; [i]Savannah River Archaeological Research Program, South Carolina Institute of Archaeology and Anthropology, University of South Carolina, New Ellenton, SC 29809; [j]Department of Anthropology, East Carolina University, Greenville, NC 27858; [k]Center for Archaeological Research, Missouri State University, Springfield, MO 65897; [l]Viejo California Associates, Joshua Tree, CA 92252; [m]Departamento de Geología y Mineralogía, Edificio U4, Instituto de Investigaciones Metalúrgicas, Universidad Michoacana de San Nicólas de Hidalgo, C. P. 58060, Morelia, Michoacán, México; [n]US Geological Survey, Menlo Park, CA 94025; [o]SRI International, Menlo Park, CA 94025; [p]Los Alamos National Laboratory, Los Alamos, NM 87545; [q]Exploration Geologist, 1016 NN, Amsterdam, The Netherlands; [r]Forschungsneutronenquelle Heinz Maier-Leibnitz, Technische Universität München, 85748 Garching, Germany; [s]Restoration Systems, LLC, Raleigh, NC 27604; [t]Kimstar Research, Fayetteville, NC 28312; [u]Faculty of Science, Charles University in Prague, 12843 Prague, Czech Republic; [v]Institute of Geology, Academy of Sciences of the Czech Republic, Public Research Institute, 16500 Prague, Czech Republic; [w]Institute for Integrated Research in Materials, Environments, and Society, California State University, Long Beach, CA 90840; [x]GeoScience Consulting, Dewey, AZ 86327; and [y]Lawrence Berkeley National Laboratory, Berkeley, CA 94720

Airbursts/impacts by a fragmented comet or asteroid have been proposed at the Younger Dryas onset (12.80 ± 0.15 ka) based on identification of an assemblage of impact-related proxies, including microspherules, nanodiamonds, and iridium. Distributed across four continents at the Younger Dryas boundary (YDB), spherule peaks have been independently confirmed in eight studies, but unconfirmed in two others, resulting in continued dispute about their occurrence, distribution, and origin. To further address this dispute and better identify YDB spherules, we present results from one of the largest spherule investigations ever undertaken regarding spherule geochemistry, morphologies, origins, and processes of formation. We investigated 18 sites across North America, Europe, and the Middle East, performing nearly 700 analyses on spherules using energy dispersive X-ray spectroscopy for geochemical analyses and scanning electron microscopy for surface microstructural characterization. Twelve locations rank among the world's premier end-Pleistocene archaeological sites, where the YDB marks a hiatus in human occupation or major changes in site use. Our results are consistent with melting of sediments to temperatures >2,200 °C by the thermal radiation and air shocks produced by passage of an extraterrestrial object through the atmosphere; they are inconsistent with volcanic, cosmic, anthropogenic, lightning, or authigenic sources. We also produced spherules from wood in the laboratory at >1,730 °C, indicating that impact-related incineration of biomass may have contributed to spherule production. At 12.8 ka, an estimated 10 million tonnes of spherules were distributed across ~50 million square kilometers, similar to well-known impact strewnfields and consistent with a major cosmic impact event.

Wittke JH, et al. (2013), Proc. Natl. Acad. Sci. USA 110 (23): E2088-E2097

Significance

We present detailed geochemical and morphological analyses of nearly 700 spherules from 18 sites in support of a major cosmic impact at the onset of the Younger Dryas episode (12.8 ka). The impact distributed ~10 million tonnes of melted spherules over 50 million square kilometers on four continents. Origins of the spherules by volcanism, anthropogenesis, authigenesis, lightning, and meteoritic ablation are rejected on geochemical and morphological grounds. The spherules closely resemble known impact materials derived from surficial sediments melted at temperatures >2,200 °C. The spherules correlate with abundances of associated melt-glass, nanodiamonds, carbon spherules, aciniform carbon, charcoal, and iridium.

Wittke JH, et al. (2013), Proc. Natl. Acad. Sci. USA 110 (23): E2088-E2097

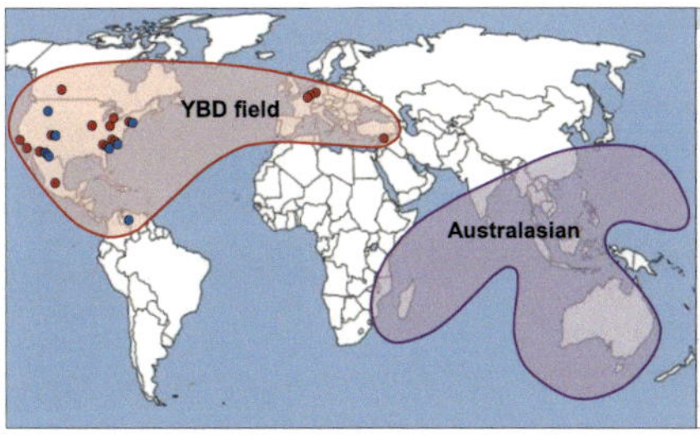

Site/Location	Details	Strat	Env	Setting
Arlington Cyn, CA	B,C,A,M	A	A	Coastal cyn
Barber Crk, NC	A	A,E	A	Relict dune
Big Eddy, MO	A	A	A	Flood-plain
Blackville, SC	--	A,E	I	Carolina bay
Blackwater, NM	B,C,A,M	A,L	A	Streambed
Chobot, ALB	B,C,A,M	A,L	A	Lakeshore
Cuitzeo, MEX	B,C,M	L	A	Lake bed
Gainey, MI	B,C,A,M	A,G	I	End-glacial
Kimbel Bay, NC	--	C,E	A	Carolina bay
Melrose, PA	--	C,G	I	End-glacial
Murray Spgs AZ	B,C,A,M	C,L	A	Streambed
Sheriden Cv, OH	B,C,A,M	C	A	Karst-cave
Talega, CA	B,A	A,C	I	Streambed
Topper, SC	A	A,C	I	River bank
Lingen, GER	B,C	E	A	Paleo-dunes
Lommel, BEL	B,CA	E	A	Paleo-dunes
Ommen, NED	B,C	E	A	Paleo-dunes
Abu Hureyra, SYR	B,C,A	A,E	I	Mound (tell)

Fig. 1. YDB impact field, based on data from 27 locations. In the YDB strewnfield (red), there are 18 YDB sites in this study (red dots; see table on *Right*). Eight independent studies have found spherules and/or scoria-like objects at nine additional sites (blue dots) located in Arizona, Montana, New Mexico, Maryland, South Carolina, Pennsylvania, Mexico, and Venezuela. The largest accepted impact strewnfield, the Australasian (purple), is shown for comparison with each strewnfield covering ~50 million square kilometers or ~10% of the planet. Table shows location of sites and lists site details (A, archeological material; B, black mat; C, charcoal; M, megafaunal remains, present either at the sampling location or in the vicinity). Also given are stratigraphic settings (Strat: A, alluvial; C, colluvial; E, eolian; G, glacial; and L, lacustrine) and relative physical stability of depositional paleoenvironments (Env: A, active, e.g., riverine, lacustrine, or eolian; I, inactive).

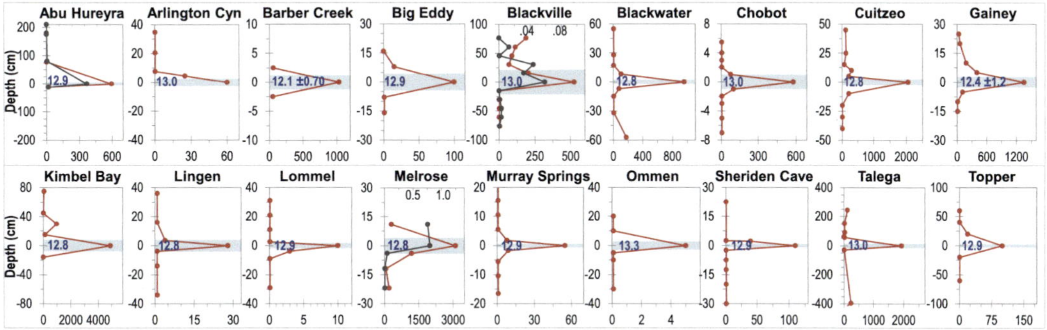

Fig. 2. Stratigraphic distribution. Abundances of spherules by site (red lines), plotted on lower *x* axis in number per kilogram relative to the YDB depth at 0 cm. SLO concentrations (black lines) are plotted on upper *x* axis in grams per kilogram (*SI Appendix*, Table S3). Thickness of sample containing YDB is indicated by blue bar. Dates for YDB layer are in blue, as determined by age-depth models (*SI Appendix*, Table S1).

Wittke JH, et al. (2013), Proc. Natl. Acad. Sci. USA 110 (23): E2088-E2097

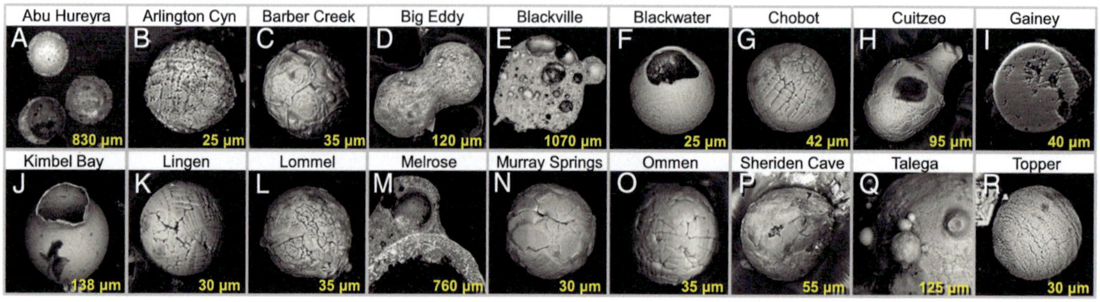

Fig. 3. YDB spherules from 18 sites. SEM images illustrate the wide variety of sizes, shapes, and microstructures of YDB spherules. Diameters are in yellow. EDS compositional percentages corresponding to the letter designations of these 36 YDB spherules are in *SI Appendix*, Table S5. Most spherules are rounded, but there are also dumbbells (*D*), bottle shapes (*H*), gourd shapes (*J*), and ovoids (*P*). Most small spherules are solid, although a few are hollow (*F* and *J*), whereas most large spherules are vesicular and/or hollow (*A*, *E*, and *M*). A large number of spherules were cross-sectioned (n = 137 EDS; *A*, *E*, *I*, and *M*); all others were analyzed whole (n = 335 EDS). Lechatelierite and flow marks (schlieren) that formed at >2,200 °C were observed in spherules from three sites (*A*, *E*, and *M*). Many large spherules display accretion with other spherules (*E*, *M*, and *Q*) and microcratering by smaller spherules (*Q*). Interior and exterior compositions of both spherules are similar, but occasionally, Fe-rich material (thin, light-colored bands) migrated or accreted to the outside of the spherule while molten (*M*). Some spherules have high percentages of TiO_2 (*B*, *L*, and *R*; averaging 42 wt%), inconsistent with anthropogenic and most cosmic origins, but consistent with impact melting of titanomagnetite or ilmenite.

Wittke JH, et al. (2013), Proc. Natl. Acad. Sci. USA 110 (23): E2088-E2097

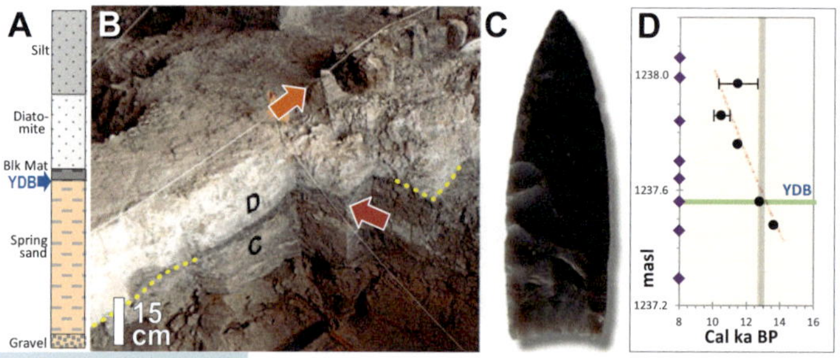

SI FIG. 4. BLACKWATER DRAW SITE (lat/long: 34.275687°N, 103.326101°W) is about 18 km SE of Clovis, New Mexico. ***A)*** Lithostratigraphic column illustrating location of YDB and black mat layer. ***B)*** Sampling site is located inside the South Bank Interpretive Center. Impact spherules were found in a 1-cm-thick dark gray YDB layer at a depth of ≈2.5 m (yellow dotted line). The YDB layer is capped by a layer of whitish-gray diatomite considered by Haynes (2008) to be the stratigraphic equivalent of the black mat. Numerous bones representing a post-Clovis Folsom-aged bison kill are shown >20 cm above the mat layer at orange arrow. ***C)*** An 11-cm-long Clovis point was found in the YDB layer nearby at the equivalent depth of the red arrow. ***D)*** Based on five conventional and AMS radiocarbon dates, we show an age-depth model (red line) using logarithmic interpolation. Errors of the older three dates are smaller than the dots. The vertical gray bar represents 12.8 ± 0.15 ka, and the purple diamonds represent sample depths. The green bar indicates the depth to the peak in impact spherules and other impact proxies in the 1-cm thick YDB layer (Firestone et al. 2007). Based on interpolation, the YDB proxy layer at 1237.55 masl dates to ≈12.8 ka.

Wittke JH, et al. (2013), Proc. Natl. Acad. Sci. USA 110 (23): E2088-E2097

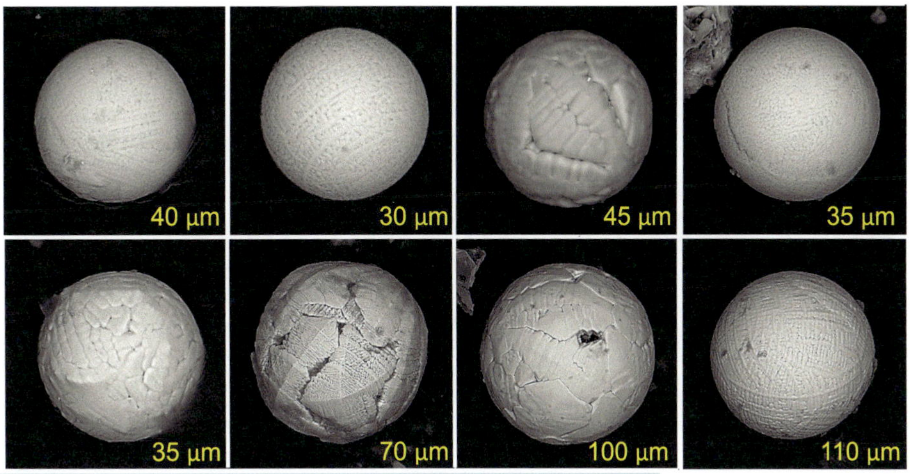

SI FIG. 18. SPHERULES FROM BLACKWATER DRAW, NEW MEXICO. SEM images display quench-melted textures indicative of rapid heating and cooling, consistent with an impact origin. Firestone et al. (2007), LeCompte et al. (2012), and this study all report finding numerous such spherules in the YDB at Blackwater Draw, whereas, Surovell et al. (2009) were unable to find any of these in the YDB layer. Spherule diameters shown in yellow.

Wittke JH, et al. (2013), Proc. Natl. Acad. Sci. USA 110 (23): E2088-E2097

SI FIG. 10. LOMMEL SITE (lat/long: 51.236231°N, 5.254686°E) is 3 km west of the town center of Lommel, Belgium. ***A)*** Lithostratigraphic sequence of eolian sands contains a thin dark, charcoal-rich layer separating tan sands above from bleached gray Usselo sands below. ***B)*** Photo of sequence shows 1-cm-thick dark, charcoal-rich layer at 48.5 cmbs (yellow dotted line) at the top of the Usselo sand. This is the YDB layer and marks the onset of the YD cooling episode. ***C)*** Two ^{14}C dates were acquired with error bars smaller than the dot size (**Table S1**). The vertical gray bar represents 12.8 ± 0.15 ka, and the purple diamonds represent sample depths. Top of Usselo sand and the ground surface are denoted by black dashed lines. The green bar shows the depth of peak in spherules in a 1-cm layer centered at 48.5 cmbs. Based on ^{14}C dates and the position of the Usselo sand, the age of the YDB layer is ≈12.8 ka.

This presentation is dedicated to the memory of "Mr. Bill", deceased Dec. 15, 2007.
Removal of Trinitite is theft of Government property and can result in fines and jail time.
William Strickfaden, Ph.D.
at Trinity Site May 7, 2004
Fellow "ant sand" enthusiast, who was instrumental to our early Trinitite work.
Trinity

Trinitite - questions?
Removal of Trinitite is theft of Government property and can result in fines and jail time.
Manhattan Project Patch
Thank You, Mr. Bill !!!
Trinity

Thom Hopen

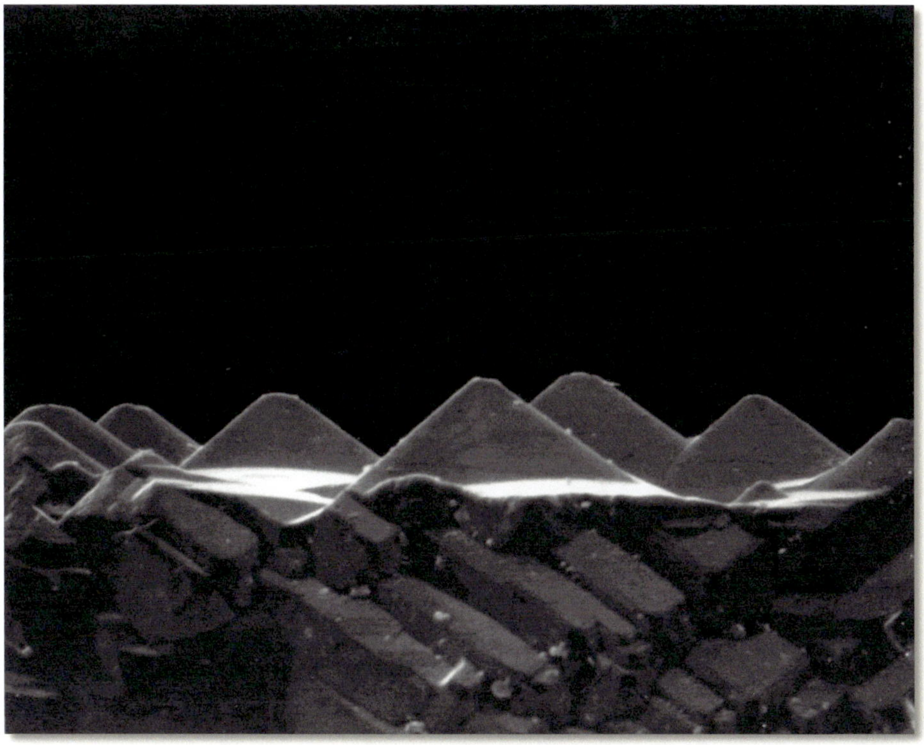

Surface of
Garnet Sand Grain

Chapel Beach,
Pictured Rocks National
Beach, Michigan

Scanning Electron
Microscope Image

Quartz Grain
from Geode

Robert Simmons

Images from a Cup of Sand

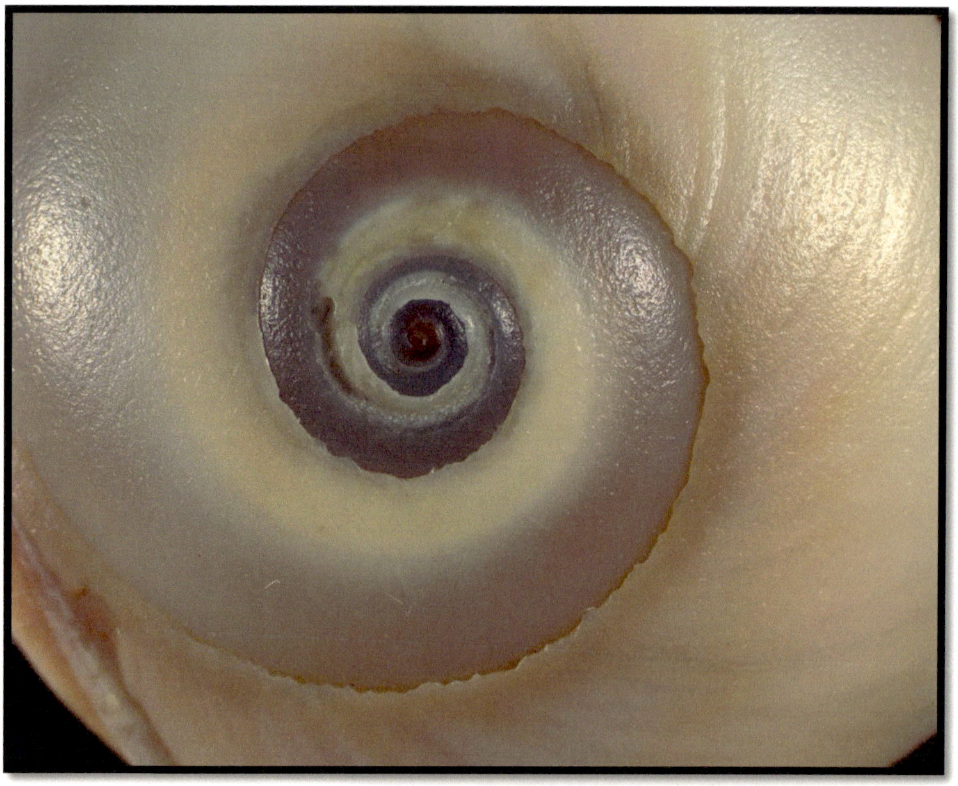

Moon Snail

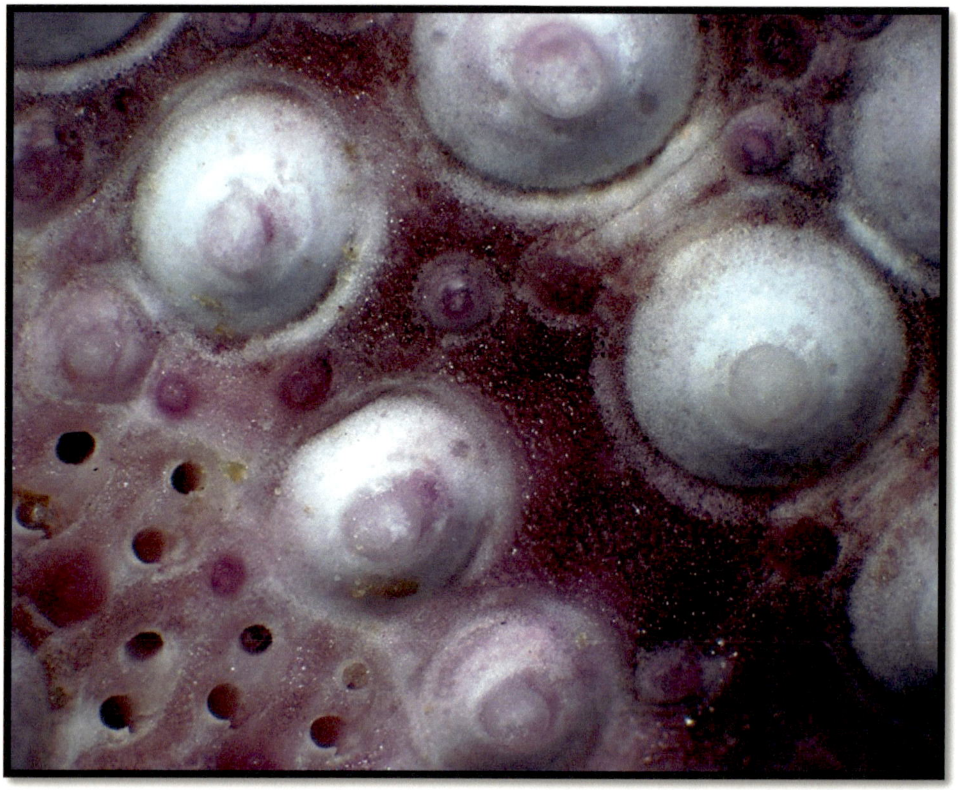

Urchin Surface

Whelk Inside

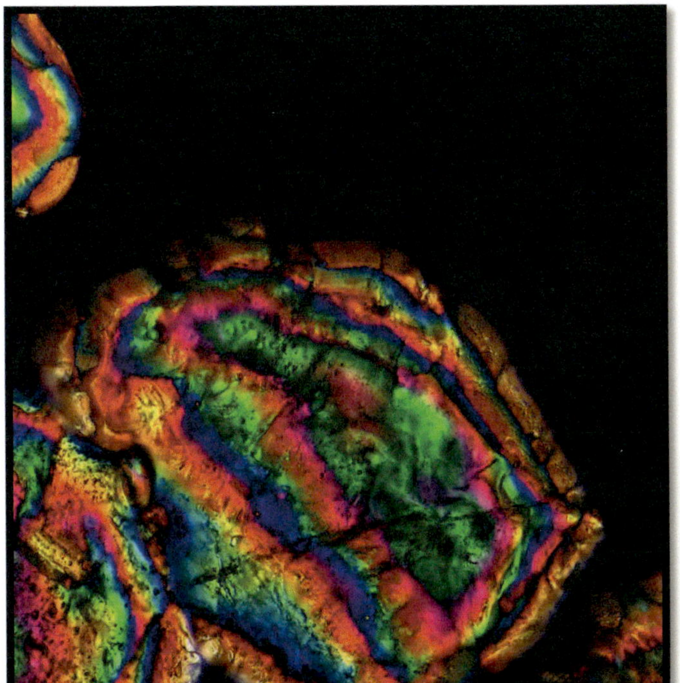

Sand Stack
Polarized Light

Sand Dollar Mouth

Worm Tube

Linda Batcha

The Accidental Sand Collector

I have been a member of the Georgia Mineral Society for about the last eleven years; mostly because my husband is an active member and has developed into an amateur geologist. My interest was more about rocks and minerals used in the making of jewelry. I also enjoyed going to the shows and on most of the field trips. Whatever I collected went into my husband's bucket of treasures for the day.

That all changed a few years ago at the Georgia Mineral Society May Show. Lori Carter was demonstrating sand. In my pessimistic way I thought to myself, “sand, really Lori, you guys will collect anything." (I may have actually said that out loud, oops). But Lori invited me to look into the microscope where she had the star sand from Japan as well as other sand samples equally as beautiful and fascinating. I was totally in awe of the intricate shapes of the sands, leading me to ask, “How can I become a sand collector?” The answer was simple: collect and share with others. I discovered that some of the advantages to collecting sand are I don't have to lug around a ten pound sledge hammer, sand is easily transported in plastic bags, and is great to trade with other collectors.

I started attending sand section meetings at the Georgia Mineral society building and sharing interesting sand I had collected.

In 2012 my husband, myself and his brother and his wife were planning a trip to New England. In my research of activities for the trip, I learned that at Plum Island, Massachusetts there was plum colored sand. I was successful in persuading the group to stop there, which was not too hard to do as my husband and his brother were already rock hounds. It was amazing, this was my premier find. I was becoming a true sand collector.

In talking with people, I have found many sand collectors, some just to remember a place they had been, others to examine it for forensic value it may provide, some have a geological interest, but for me, it is the beauty that has me hooked.

Tom Batcha

The Iron Furnaces of Upper Stamp Creek

and the Role that Sand Played in Their Operation

The Iron Industry at Upper Stamp Creek

The iron industry in Bartow County dates back to the early days of the county's history. In 1837, the ironmaster Jacob Stroup had purchased land on Lower Stamp Creek in the 21st district and soon after constructed the first iron furnace. Jacob was joined later by his son Moses who eventually took the operation. By the 1850's there were eight furnaces in operation in the county with the center of the industry around the town of Etowah. Much of this expansion came with help from a politician turned ironmaster named Mark Anthony Cooper who had met up with Moses Stroup and realized the potential and formed a partnership with him. Cooper later bought Stroup's share and continued to expand the operation. By the time of the Civil War the iron operation at Etowah would consist of three iron furnaces, a rolling mill, a nail factory, warehouses, hotel, store, and a three-story flour mill. In addition, there were numerous furnaces across the county producing iron that was sold to the rolling mill in Etowah.

One of the thriving areas of the county during this time was on the upper branches of Stamp Creek located southeast of the current day town of White. During the boom years, there were four iron furnaces, a forge, and a merchant mill in operation around the community named Stamp Creek. Much of the success in this area can be attributed to John W. Lewis. He was a doctor, wealthy land owner, preacher, and iron master. He served in the South Carolina legislature in 1830 and 1831 before moving to Georgia in 1840. In 1857, he was appointed the superintendent of the Western and Atlantis railroad by Governor Joseph E Brown. Lewis did much to improve the roads around Stamp Creek and to connect it with the more popular parts of the county. As a wealthy land owner, he owned the merchant mill (Lewis Mills), two of the iron furnaces (Lewis Furnace and "the New Stack"), 50% interest in the Pool Furnace and Forge, and was the original land owner of the land for the Union Furnace property. The remains of the four furnaces can still be seen today within the boundaries of the Pine Log Wildlife Management Area.

Overview of an Iron Furnace Operation

The iron furnace has three major parts in its operation: The Stone Stack, the Charging House, and the Casting Shed.

In the charging house, raw materials needed for the making of iron are fed into the top of the stone stack. Those materials are iron ore, charcoal (this was later replaced by coke), and limestone that was used as flux. These materials were stockpiled in the area behind the charging house and were loaded into the stack in rotating charges. When locating a site for an iron furnace in the early to mid-1800s, it was important that these resources were available in the close proximity. Two other prerequisites for the site were a creek to harvest water power to run an air pump apparatus and a hillside to build the stone stack against where the charging house would operate as a bridge between the hilltop and the top of the stone stack.

The Stone Stack is at the core of the operation. Here the charcoal would fuel the furnace to the temperature of the melting point of iron. This process requires that air is pumped into the furnace by way of a funnel known as a tuyere. As the iron would become liquid, it would accumulate at the bottom of the furnace known as the bosh. The impurities from the iron ore are mixed with the limestone which acts as a flux and forms the waste byproduct known as slag. The slag is lighter then the iron and is tapped off the top with a slag ladle.

Once the liquid iron reaches the right level in the bosh, the furnace is tapped and the liquid iron flows out of the furnace into the area known as the casting house.

In the casting house, the iron would be tapped once it which its melting point. The floor of the casting house was covered with sand that had channels formed so that the liquid iron would flow from the furnace down a main channel known as a sow. In this sow channel the flow of the iron would be diverted by a shut plate into side channels that emptied into molds patterned in the sand known as pigs. These molds were about three feet long and would hold about 100 pounds of iron when cooled. Once these pigs were full, the shut plate would be removed and the iron would flow further down the sow to the next shut plate where the next series of pigs would fill and the cycle continued. The terms come from the resemblance of a mother sow feeding her piglets which gives us the name pig iron.

How Sand was used at an Iron Furnace Site

There are a number of uses for sand at the iron furnace.

Sand used in the Casting Floor:

The main use of sand in iron making was in the casting shed. The sand used in the casting shed had a high silica content and was normally white to yellow in color. The entire floor of the casting shed would be covered with the sand. Along with the channels created for the iron to flow into the pig iron molds, there were also individual molds created that were filled by a ladle carried by a team of men. These molds were made in various shapes and formed iron objects such as large pots and tools needed by the local population.

Sand used in the Stone Stack:

The stone stack was made of two different walls. The outer wall was made of large stone blocks of sandstone and an inner wall was made up of a chimney lining. In between these two walls was a buffer area that was filled with sand. It has also been found that local sand was used in the making of mortar for the bricks in the chimney lining.

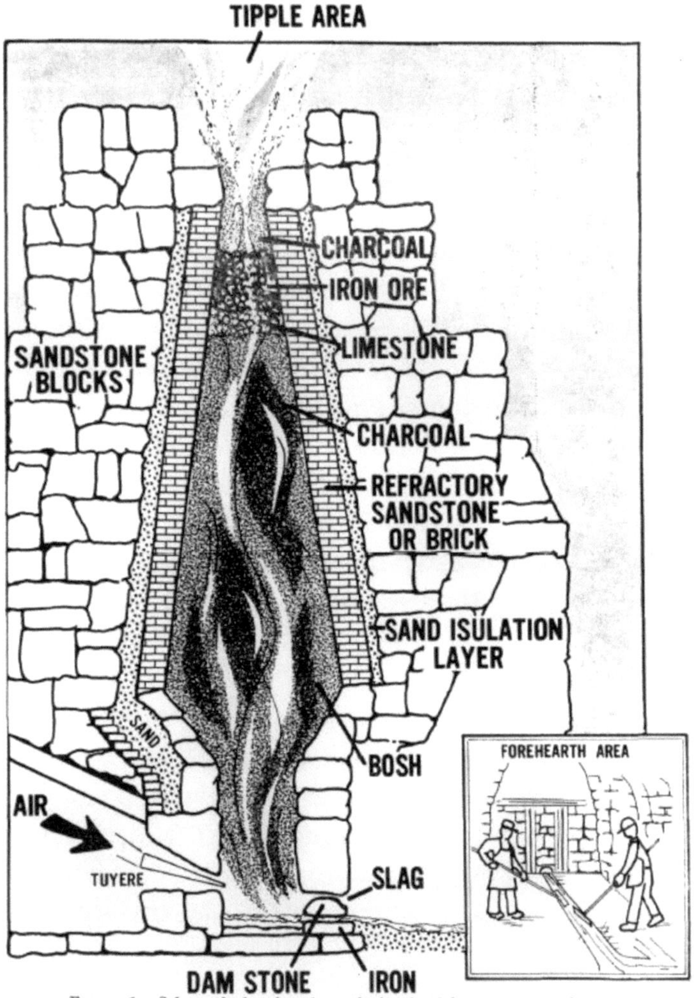

FIGURE 1. Schematic drawing of a typical early 19th century blast furnace.

Source:
Preliminary Archaeological Examination of Ohio's First Blast Furnace: The Eaton (Hopewell), by John R. White, The Ohio Journal of Science. v80, n2 (March, 1980) pages 53

Sand Research at Iron Furnace Sites

Recently, there have been a number of iron furnace sites that have been the topic of industrial archeology projects. Found below are two different studies of iron furnace sites that provide the reader with some insight to the study of sand in the iron making process.

Preliminary Archaeological Examination of Ohio's First Blast Furnace: The Eaton (Hopewell), by John R. White, The Ohio Journal of Science. v80, n2 (March, 1980) page 54

> *The inner chimney lining was made from shaped blocks of refractory sandstone that were cemented with a hard mortar colored to a brick-red. Examination of this mortar showed a constituency quite similar to that of the samples of sand and soil analyzed from the site (table 1). In a laboratory experiment, red sand from the hearth opening (Sample 2) was mixed with water and baked. This process resulted in the creation of a friable concretion not as hard as but not very different from the color and texture of our mortar sample. This finding leads us to believe that the furnace builders expeditiously welded their sandstone blocks together with mortar made from local sand or mud and water. The intense heat and pressure produced by the furnace operation hardened and colored the mortar.*
>
> *FURNACE WALL*
> *The outer furnace wall was composed of large hand-chiseled blocks of native sandstone, some weighing several hundred pounds. The insulating space between the heavy outer wall and the refractory inner chimney was filled with sand that was subsequently oxidized to a bright red-orange by the intense heat generated by the furnace. This sand, though dramatically different in color, proved to be quantitatively and qualitatively identical to samples taken from other areas of the site (table 1). These sands were local in origin and resulted from weathering and erosion of the contiguous sandstone cliff.*

Archaeological Investigation of the Casting Area at Site 1Tu604, The Tannehill Furnace Compound, by Beverly S. Curry, University of Alabama (October, 1995) Page 33

> *CASTING SHED*
> *The area from 1.5m (4.9 ft) in front of the Double Furnace to the line of the stone wall was shovel skimmed. After the rocky clay surface was manually removed, it became apparent that the entire underlying surface of the area within the stone wall boundary is composed of the yellow and brown sand, comprising the casting floor. The casting floor is a 6 m (19.7 ft) square area of sand with numerous pebble-sized iron pieces concentrated on the northern side. The first layer of sand is yellow. It is clean and sterile. Beneath this is a layer of brownish sand which is contaminated by spew.*

The Furnace Stacks of Upper Stamp Creek

Lewis Furnace
Built 1847
Land Lot 151, 22nd District, 2nd Section
Names used while in Operation:
1859: Lewis Cold-blast Charcoal Furnace (Lesley)
1864: Earle Furnace (County Deeds Book P)
1869: J. E. Brown & Co. Iron Works (County Map)
1876: Bear Mountain Furnaces (Iron & Steel Directory)
1886: Oak Grove Furnace, Brown & Thomas (I&S Directory)

Union Furnace / Ford Fire Eater - Diamond Furnace
Built 1852
Land Lot 114, 22nd District, 2nd Section
Names used while in Operation:
1859: Union Cold-blast Charcoal Furnace (Lesley)
1862: Novelty Ironworks (County Deeds Book P)
1864: Withers Factory (Atlanta Campaign Map)
1876: Ward's Diamond Furnace (I&S Directory)
1886: Diamond Furnace, W.P. Ward (I&S Directory)

Pool Furnace
Built 1855
Land Lot 66, 22nd District, 2nd Section
Names used while in Operation:
1859: Pool Cold-blast Charcoal Furnace (Lesley)
1864: Pools Factory (Atlanta Campaign Map)
1869: Pools Iron Works (County Map)

"The New Stack"
Built 1863
Land Lot 150, 22nd District, 2nd Section
Names used while in Operation:
1864: New Furnace - unfinished (County Deeds Book P)
1876: Bear Mountain Furnaces (I&S Directory)
1886: The New Stack, Brown & Thomas (I&S Directory)

Kate Clover, Carol Hopper Brill, and Leo Kenney

Sands and Their Stories

Ushuaia-Tierra Del Fuego, Argentina

Charles Darwin, in 1838, was the first naturalist to write about Tierra del Fuego, the southernmost tip of South America. He noted that the rocks around Ushuaia included metamorphic schists, phyllites, gneiss and granite. He also noted the change of direction of the Andean Cordillera from a N/S direction at the Straits of Magellan to an E/W direction south of the Beagle Channel. How right he was! Today scientists studying the complex geology of this region recognize geologic processes including volcanism, stretching, compression and faulting where the oceanic Nazca plate wedges itself under the South American continental plate.

Sediments in this sample derived from the decomposition of rocks along the shoreline. Other grains no doubt originated in the Andes and made their way to the shore via the many rivers.

Mar Del Plata, Argentina

Argentina's premier beach resort, dating back to the late 1880s, Mar del Plata is located on the Atlantic Ocean 250 miles southeast of Buenos Aires.

In addition to miles of "golden" sands, the shoreline boasts one of the oldest rock formations on the planet. The Tandilia System stretches across the province and enters the ocean near Mar del Plata. This two billion year old formation includes igneous and metamorphic rocks once part of the supercontinent Rodinia that predates Pangaea.

Mineral grains include clear quartz and reddish feldspar. Also abundant are highly polished white and tan mollusc shell fragments including the multi-layered brown chunk (lower center). Thin, translucent chips may come from intertidal coquina bivalves like *Donax*. The blue-black grain may be from a *Mytilus* mussel.

Nosy Be, Madagascar

The island of Nosy Be lies five miles off Madagascar, facing the Mozambique Channel. Despite the island's ancient volcanic origins, the sand from this tropical Indian Ocean beach reflects the active reef that thrives nearshore and helps make Nosy Be this nation's largest and busiest tourist resort.

This sample features bits of *Halimeda* coralline algae, like the irregular chalky oval piece along the top near 1 o'clock. Forams are represented by the tan disc *Marginipora* and the red *Homotrema* (left). Denser white chunks are likely coral. Whole snail shells include white and brown micromolluscs, a transparent shell, and the red and white top snail. A white bivalve shell is filled with smaller debris. A rose-colored urchin spine sits between the top shell and a white foram.

Cape Lardier, Cote d'Azur, France

On France's famous Mediterranean Cote d'Azur, Cap Lardier lies south of the renowned Riviera resort Saint Tropez. The Cape is a conservation area with spectacular views of rocky slopes and pocket beaches.

Biogenous material dominates this sample. Coralline algae include the red articulated branch (top center) as well as the large chalky lump (near center). The red chips are the foram *Miniacina miniacea.*

Micromolluscs include snail shells and a keyhole limpet (left middle). The bumpy green chip and the perforated fragment (center bottom) are urchin tests; the mauve rod is an urchin spine. The lacey vertical column (right side) and white chunk (far left) are probably calcified bryozoans. The P-shaped fragment (center) is a vermetid snail. The perforated white fragment (bottom left) may be coral or shell.

Auke Bay, Juneau, Alaska

This sample from Juneau, Alaska is a "surf and turf" sand with both rock grains from the land and marine-life fragments from the sea.

The Mendenhall Glacier, one of North America's largest ice fields, drains into Auke Bay, north of Juneau. As the glacier journeys to the sea, it scours and transports rocks from the jagged peaks and valleys, including orange/brown granite and clear quartz. In addition, this beach sand includes fragments eroded from the shoreline. These land-based grains compose the "turf" components of the beach.

The "surf" grains in this sand are shell fragments of marine invertebrate species that inhabit the local waters: blue mussels, white fragments likely from oysters and barnacles; and the striated green rod is an urchin spine.

Jeffrey's Bay, South Africa, East Cape (Indian Ocean)

Jeffreys Bay, on the Eastern Cape of Africa, is famous for its surfing and pristine beaches. The landscape of the area is dominated by two mountain ranges with high peaks and rocks that date back 800 million years to the Pre-Cambrian. Quartz-rich sandstones in colors from pink to clear and brownish dominate the landscape. Transported over millennia down the mountain streams, quartz makes up the bulk of the grains in this sample.

Colorful biogenic contributions from the ocean include shells and skeletal shards from invertebrates and algae, including: a red nerite snail; numerous bivalve fragments; green, purple and bi-color urchin spines; yellowish rods of coralline algae and numerous ribbed shell fragments. The white disk with a hole (left of the red nerite) is likely a foram.

Miller's Point, Cape Town, Western Cape, South Africa

Miller's Point is located south of Cape Town near the Cape of Good Hope. It is known as a dive site with dense kelp forests and a variety of invertebrates, including sea urchins that inhabit the shallower waters. Divers note they often see urchins clinging to kelp stems and blades with their many tubefeet while feasting on the kelp. Sea urchins have one of the most complex feeding structures in the animal kingdom. Called the Aristotle's lantern, it contains five teeth that urchins use to rasp algal tissues. Both the orange and purple rods are urchin spines. The red and white chips in the center are mollusc shell fragments.

The area rocks are granites and sandstones, both of which contribute the quartz grains in this sand.

Cape Town, South Africa

Cape Town, just west of the Cape of Good Hope—Africa's southernmost tip—is known for its diverse coastlines: massive rocky headlands, long stretches of surf-swept sandy beaches and rocky intertidal areas dotted with pools teaming with marine life.

This sand illustrates some of the diverse invertebrate life found in those rich waters. The transparent rods are sponge spicules, calcareous or siliceous (silicate) structures that are embedded in a sponge's tissues. Spicules may interlock or be fused together in a scaffolding that supports the fibrous protein mesh that makes sponges ... spongy. The grooved rods are sea urchin spines. Bivalves include the large dark brown and light brown shell fragments, as well as the tiny white clams and brown bubble snail shell near center at 1 o'clock.

Hummingbird Cay, Great Exuma, Bahamas

Here is an example of sand that is locally produced and is not the legacy of glaciers or ancient mountain ranges. The tiny white "eggs" in this sample are ooliths, from the Greek word meaning "egg stone." (Larger than 2mm, they are called pisoids.) They form in warm, shallow tropical waters by chemical precipitation of dissolved calcium carbonate around a nucleus. As the waves and currents roll these grains around, carbonate minerals coat the surface, and the eggs continue to increase in size, like snowballs, until they are too big to roll. In cross section, they show concentric rings.

This sample also includes a coral fragment (lower left), white forams (ear-shape at 9 o'clock and ramshorn at 6 o'clock), red forams (*Homotrema rubrum*) and thin green fishing line.

Bretignolles sur Mer, France

Bretignolles sur Mer is on the Atlantic coast of France. This concentration of pink and red garnets, black iron-rich minerals (mostly magnetite), green epidote, and milky to clear quartz makes this a placer deposit. Placer mineral deposits show up on beaches where the waves, wind and tides segregate the heavy minerals from the less dense ones and form colorful lines and dark swaths on the less dense grains, often quartz and feldspar.

The geology of this region is complex and dates to about 280 million years ago with the collision of two continents, mountain building, and the formation of the supercontinent of Pangaea. The mineral grains in this sand are derived from the roots of those ancient mountains.

Valentino Beach, Mazatlan, Mexico

A tourist destination on Mexico's Pacific coast, Mazatlan is located at the mouth of the Gulf of California. Characterized by high biodiversity, this region supports diverse coastal ecosystems and thousands of marine species.

Ranging from translucent white to orange and brown, the glassy sand grains are primarily from calcium carbonate mollusc shells, including the tooth-shaped fragment at 4 o'clock. A narrow cone-shaped snail at about 7 o'clock is an example of a micromollusc—a mollusc species that doesn't grow any larger than a grain of rice.

The rugged Sierra Madre Occidental mountains along this coastline are comprised of rocks that chronicle volcanism and pyroclastic activity. The cement-like grains, called ignimbrites, are cemented pumice and ash. The black grain is probably basalt.

I-35W Bridge, dredge, Mississippi River, Minneapolis, MN

This plain brown sand includes sediments the Mississippi River continuously hauls downstream from its headwaters in Northern Minnesota, some 230 miles before reaching Minneapolis.

The sand grains originated as both intrusive and extrusive igneous rocks and other ancient rocks that the Ice Age glaciers ground up and transported some 2 million to 10,000 years ago. Grains derived from granite include clear quartz and pink feldspar; grains from the northern part of the state include magnetite, ilmenite, red agates and lithics. The sample also includes evidence of the industrial city: chunks of bubbly slag, charcoal, and a piece of yellow road-strip paint embedded with some reflective glass beads. The hollowed orb is likely from a sand blasting operation. The semi-circle bottom left is part of a freshwater mollusc.

Lake Superior, Knife River, MN

With the assistance of the glaciers, this rocky looking sand eroded from their iron-rich volcanic rocks common in northern Minnesota. Near the Lake Superior shoreline, the rivers have cut deep gorges and carry sediments from the uplands to the lake. The bedrock along the shore also erodes and contributes to this beach "sand."

Many of these grains have a red hue—a clue they contain some iron. Magnetite, an iron mineral, is the black grain in this sample.

The Lake Superior region has a rich mineral history. Beginning in the late 1800's, miners extracted massive amounts of copper from Michigan's Keweenaw Peninsula and iron from northern Minnesota, Wisconsin and Michigan. Some iron mining continues and there is ongoing exploration for other minerals as well.

Glass Beach, Ft. Bragg, CA

Fort Bragg Beach is known as the Glass Sand Beach because glass is an abundant grain material amid the rock grains. Between 1949 and the 1960s, trash, including many glass bottles, was thrown over the cliff. Today in amongst the rock and urchin spines, the beach contains millions of pieces of glass of many colors and maybe a fragment of an old dinner plate.

Fast-forward to 2007. Coastal engineers in Fort Lauderdale, Florida are taking a lesson from the Fort Bragg Glass Sand Beach. They are grinding up clear glass bottles, polishing the grains, and using this material to replenish beach sand lost to erosion and severe storms.

Juodkrante Beach, Curonian Spit, Lithuania

The Curonian Spit on the Baltic Sea is known as the "Amber Coast." This sandy line of massive dunes has migrated over the years burying 4000 year-old hearths and many medieval settlements. Archaeologists have found carved amber jewelry and sculptures at these settlements and at a small embayment in the lagoon of Juodkrante. The Amber Bay is a prime amber-mining locality.

Amber, fossilized sap from 15-20 million year old tree, is found on rivers, fields and on beaches; it continues to be integral to the local economy and regional identity.

The three orange grains are amber. The black "conglomerate" is an oil blob imbedded with other grains—probably a remnant from an oil platform spill in the Baltic

Carol Hopper Brill

Skeletons in the Sand: Sources of Biogenous Sand Grains

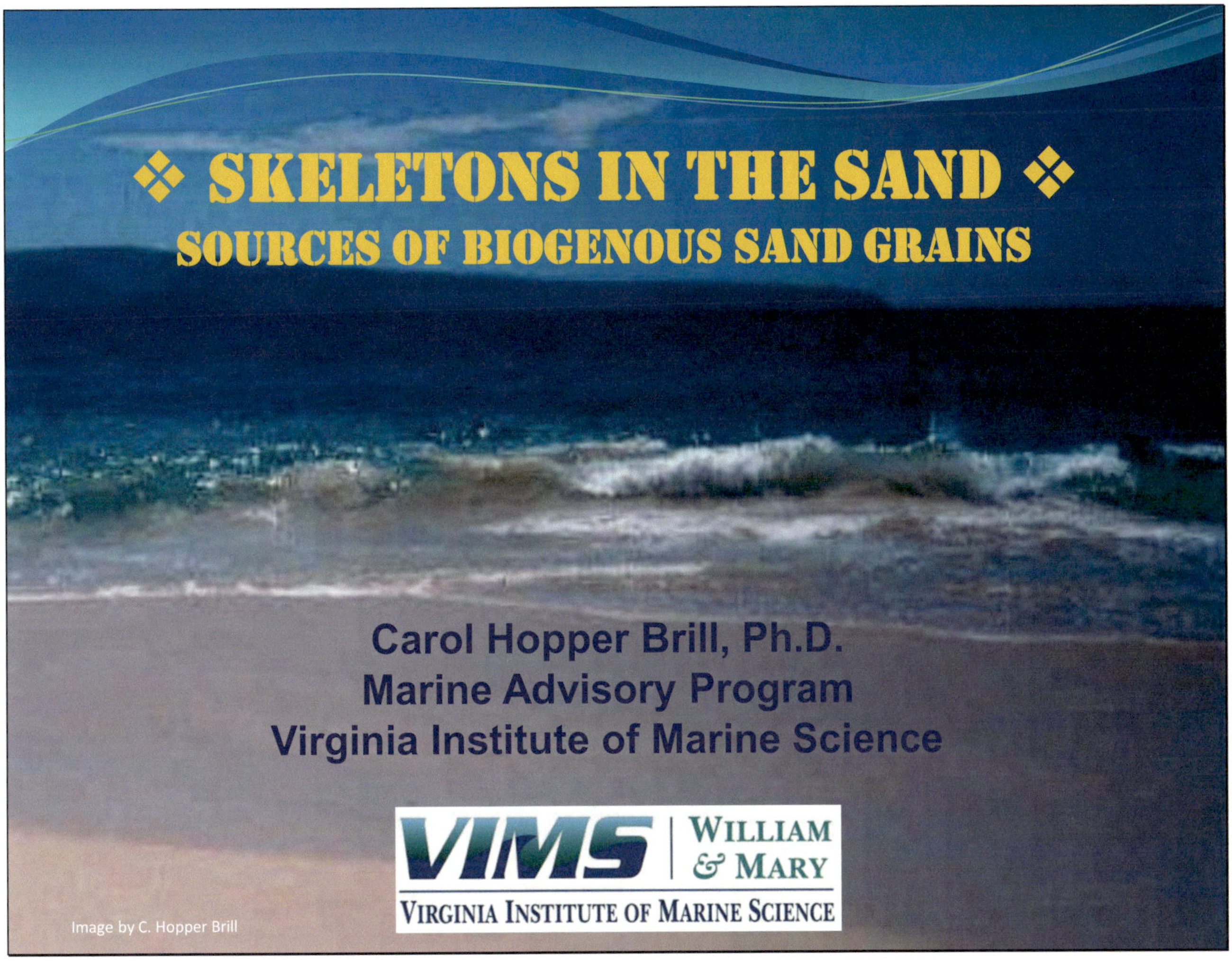

Carol Hopper Brill, Ph.D., Marine Advisory Program,
Virginia institute of Marine Science, College of William & Mary
Gloucester Point, Virginia

Taking a closer look at sand tells us about the beach under our feet, about the mountains that may be hundreds of miles away, or about the marine life living along the shore. Do a bit of forensic beachcombing and you can identify some of the skeletons in the sand. This presentation introduces some of the most common biological sources of sand.

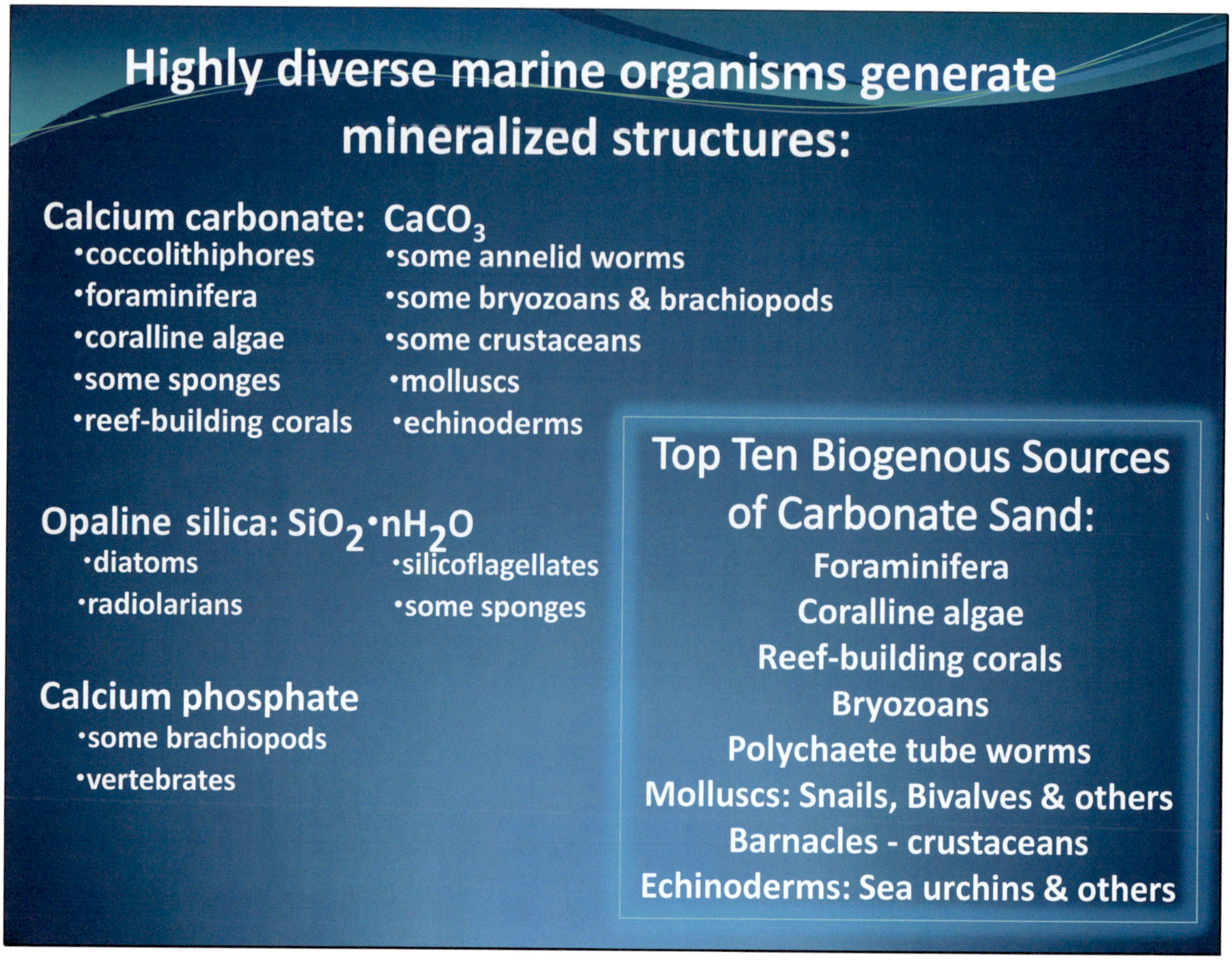

Diverse marine organisms generate mineral compounds for skeletal, support and defense functions through processes of biomineralization.

•Calcium carbonate is the mineral found in the skeletal frameworks of a wide variety of protists, algae and invertebrate animals. These diverse organisms take up calcium and carbonate ions from the surrounding seawater and through cellular processes deposit them as mineral $CaCO_3$. A protein matrix is often secreted that serves as a framework for mineral deposition.

•Opaline silica (SiO2•nH_20) is the skeletal compound for some important protist groups (most familiar are diatoms and radiolarians). And, two groups of sponges have silicate skeletal components.

•Calcium phosphate is found in fewer kinds of marine organisms, but vertebrate bone is made of this compound.

This presentation focuses on about ten particularly common biogenous sources of calcium carbonate sand: Foraminifera; coralline algae; reef-building corals; bryozoans; polychaete tube worms; molluscs, including snails, bivalves & others; barnacles, which are crustaceans; and echinoderms, including sea urchins and their relatives.

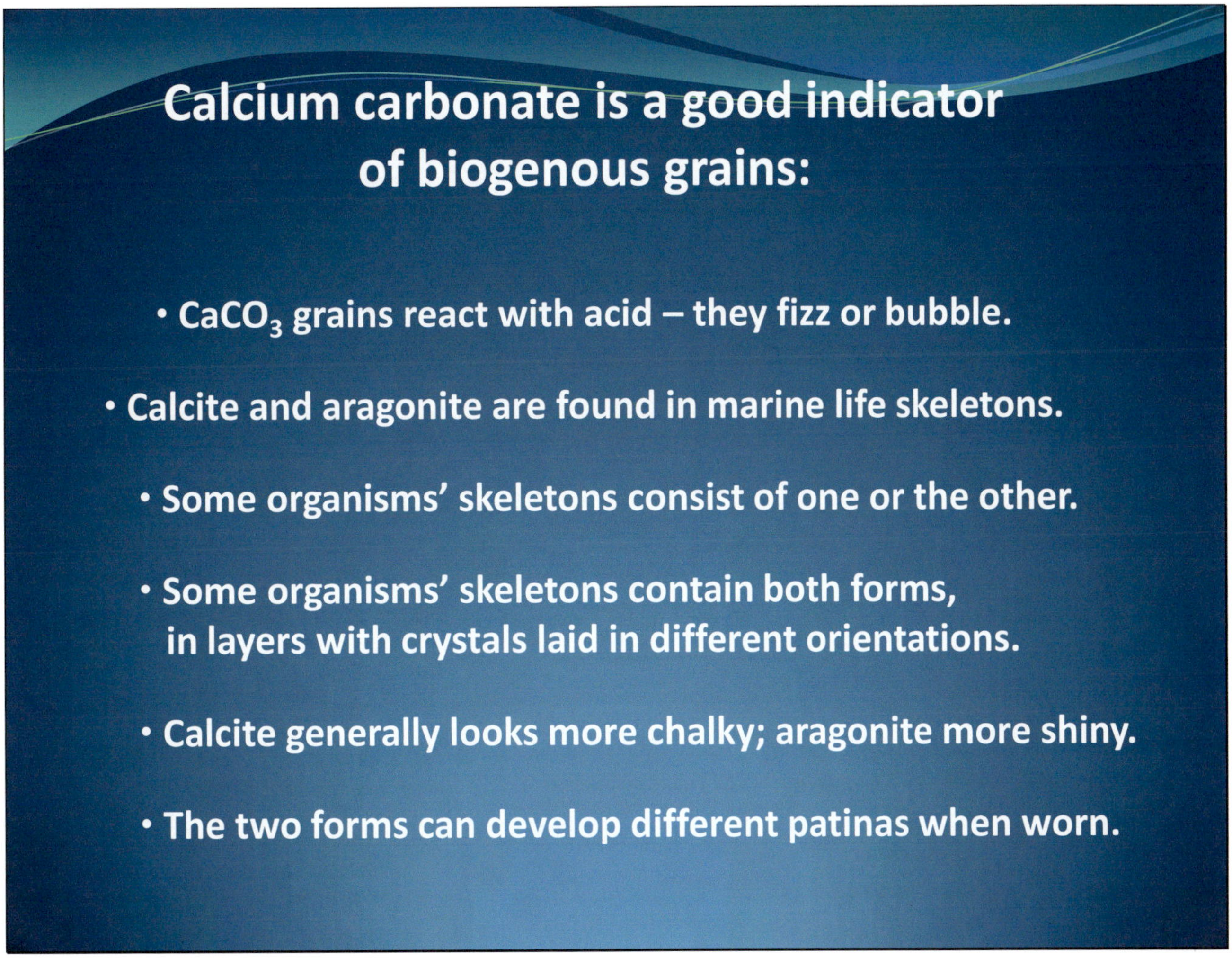

So, calcium carbonate is a good indicator of the presence of biogenous material in sand. And, it can be easily confirmed using weak acetic acid – household vinegar. Drop a few grains of calcium carbonate shell material in vinegar and you'll observe the development of bubbles. As the acid dissolves the shell, carbon dioxide gas (CO_2) is released.

Of the three forms of calcium carbonate, the two produced by marine organisms - calcite and aragonite - both react with weak acid. The presence of one or both forms can be diagnostic in identifying the source of shell material in sand. Calcified structures of some organisms are composed of either calcite or aragonite exclusively. The shells of others contain layers of both calcite and aragonite. In general, calcite looks more chalky, aragonite more shiny. They may develop different patinas as they weather, but can be differentiated by their microscopic crystal structures or by using etching and staining techniques. These two carbonate minerals can be differentiated from the third, dolomite, which does not easily react to acid and has a crystal structure different from either calcite or aragonite.

Silicate skeletons and most mineral grains will not react in the acid.

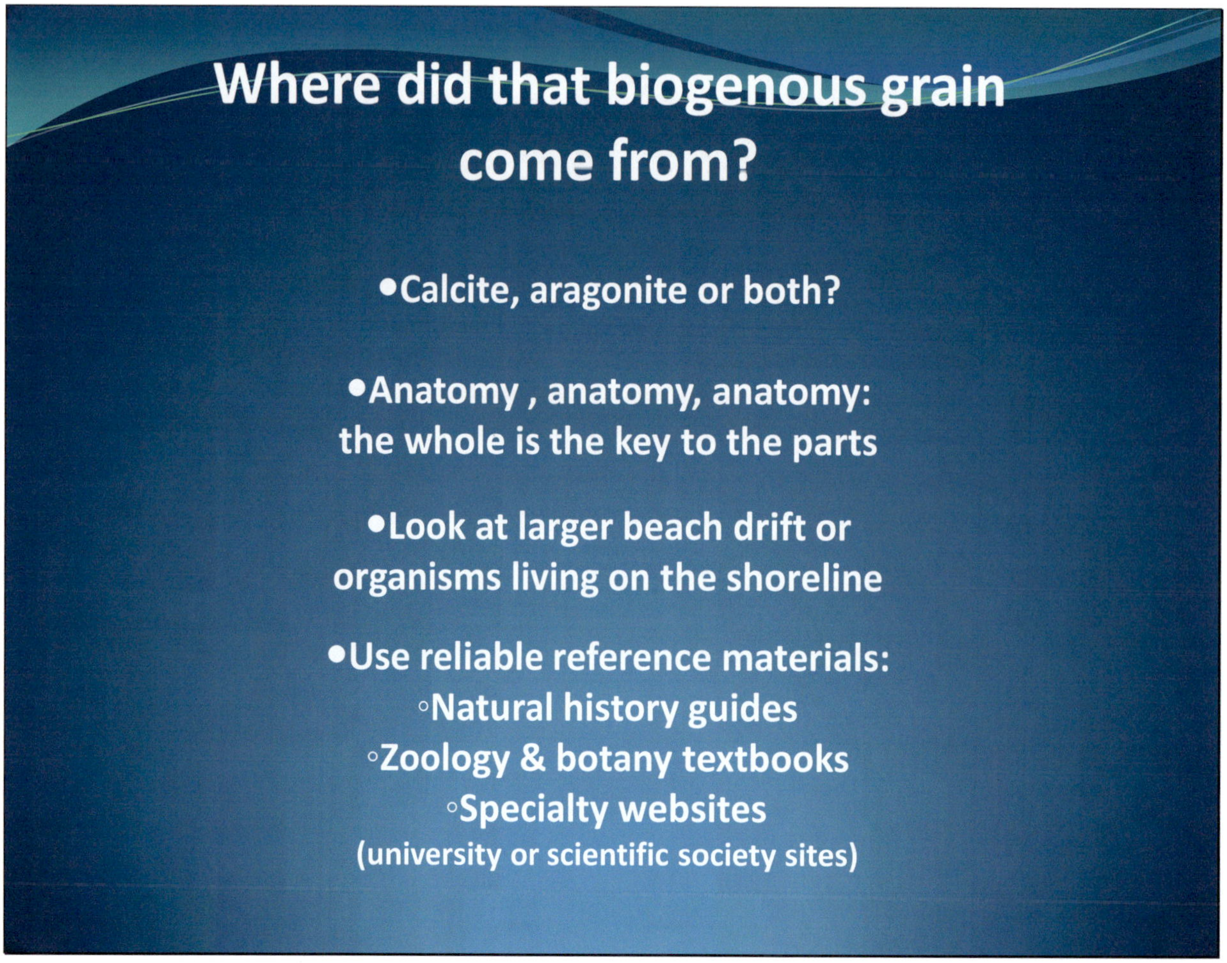

Once you know your sand contains biogenous components, how do you know what organisms they came from?

•Calcite, aragonite or both? Sometimes the surface patina provides clues. Calcite is general chalky looking; aragonite more polished.

•Anatomy: the whole is the key to the parts. To hone your diagnostic skills and build search images, you need to look at whole organisms and match up the anatomy of different structures. Start with a good marine life identification key for the area your sand comes from.

•Look around your collecting site for larger pieces of beach drift or for living organisms on the shoreline or in the sand. These provide clues to the smaller pieces found in the sand.

•Use reliable reference materials: A well-illustrated invertebrate zoology textbook will be helpful since the skeletons and shells of many invertebrate groups contribute to sands. And, don't neglect marine algae; calcified forms can be major sand contributors, so an illustrated key to marine plants or botany text can be helpful. There are also specialty websites that offer images of marine life. Be sure to use those that are properly vetted by experts (such as university or scientific society sites) so that you avoid misidentifications that are rampant on the internet image collections. This brief overview shares some key features and clues I look for when I'm identifying the sources of individual biogenous sand grains.

Foraminifera, forams for short, are single-celled protozoans related to the amoeba. They secrete a chambered calcium carbonate test (shell) that increases in size as the cell grows. Cell cytoplasm extends out through small pores (foramina) for movement and to capture food particles. The distinctive test forms are diagnostic and detailed websites provide images and keys.

Some forams drift in the planktonic, others live on the seafloor or surfaces of other bottom dwellers; the ones pictured here are larger, bottom-dwelling species. Some tropical forms, like the large discs seen in the top center and left, may reach the diameter of a nickel (2 cm). These species often are found to have a symbiotic algae living within their cytoplasm. The chemistry of algal photosynthesis facilitates greater calcium carbonate deposition. Attached red forams, seen lower left, are quite distinctive and almost look like miniature corals. Their red color can make some sands look pink. Tropical "star sand" forms are known world-wide for their shape. For an example of an online photo key to forams see: http://www.foraminifera.eu/AtoC.html

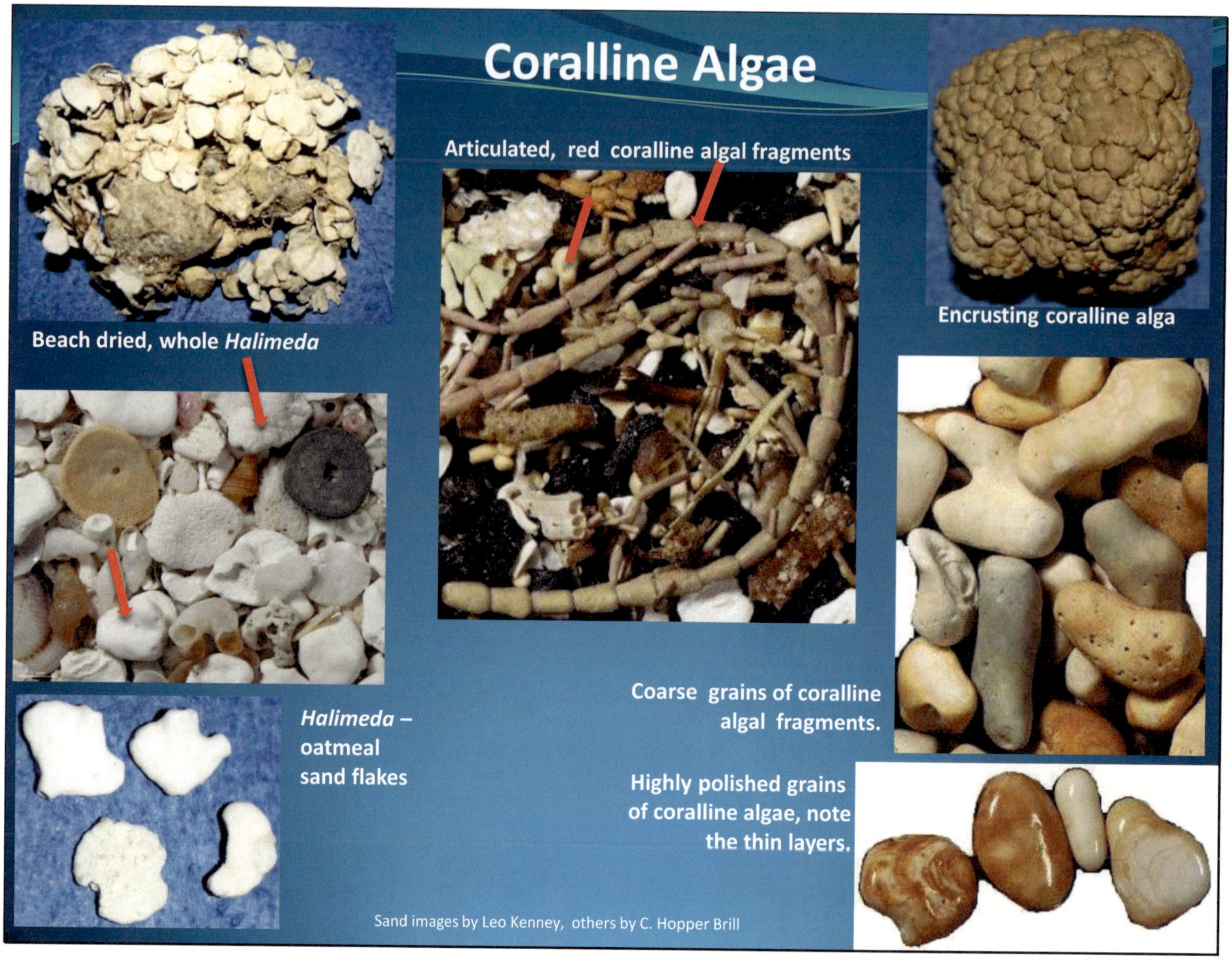

Animals are not alone in secreting mineralized structures for support or protection. Some kinds of green algae (Chlorophyta) and red algae (Rhodophyta) are calcified. The added reinforcement provides strength for survival in wave-impact zones and protection from marine herbivores.

Of the calcified greens, one of the most common is *Halimeda*, a tropical form with erect blades of articulated, disc-shaped sections. When the alga dies, the calcite segments come apart into chalky white flakes (up to ~1/4" across). In some locations, these grains are dominant contributors to the sediment, forming "oatmeal sands."

Among the red algae, there are both erect, articulated growth forms, as well as encrusting forms that grow on a variety of surfaces from rocks to other organisms. Red coralline algae occur from tropical to polar seas. Some encrusting species form knobby heads that might be mistaken for corals, but they look different at the surface and in cross section. Coralline algae lack the distinct surface calyxes (polyp cups) seen in corals and have very thin growth layers. When fragments of coralline algae weather, they often retain a chalky finish that may range from white to rusty orange. Grains that have been polished take on a shiny finish, but may still show the thin growth layers.

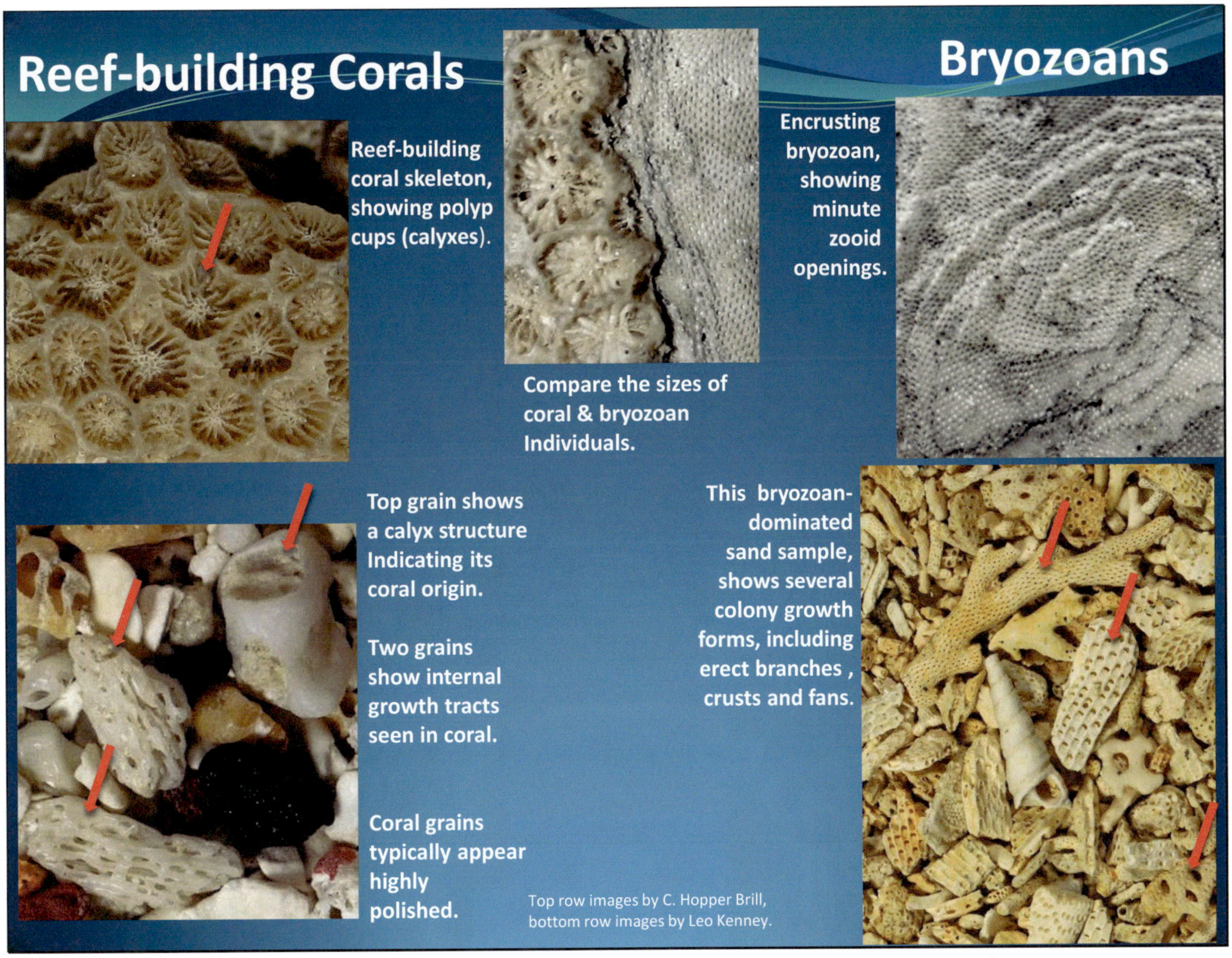

Two colony building animal groups – corals and bryozoans - contribute to sand and can be confusing to tell apart. Skeletal size and structure holds the key to differentiating them.

Reef-building corals are Cnidarians, skeleton-building relatives of sea anemones and sea jellies. Individual coral polyps have a central mouth surrounded by stinging tentacles. Most reef-building forms are colonial, the interconnected polyps forming a thin living skin over the aragonite skeleton they secrete. Each polyp has a supportive cup (calyx) into which it can contract – the presence and size of these cups is one diagnostic for identifying coral fragments in sand. Another feature unique to corals is the ladder-like growth tracts in the skeleton formed as the colony grows. Finally, the aragonite wears to a high, glassy polish. Reef-building corals are limited to warm water locales, but some corals do exist in cooler waters.

Bryozoans, or moss-animals, belong to an animal group that is found from tropics to poles. These colonial organisms have skeletons made of flexible protein which, in some species, is reinforced with calcium carbonate. Growth forms include erect branching or fan-like colonies, as well as crusts on other surfaces, from rocks to seaweeds. The microscopic individuals, called zooids, live inside compartments visible as small openings in the surface. The chalky skeletons of bryozoans are thinner, less dense and more brittle than corals and are less likely to obtain a high, glassy polish.

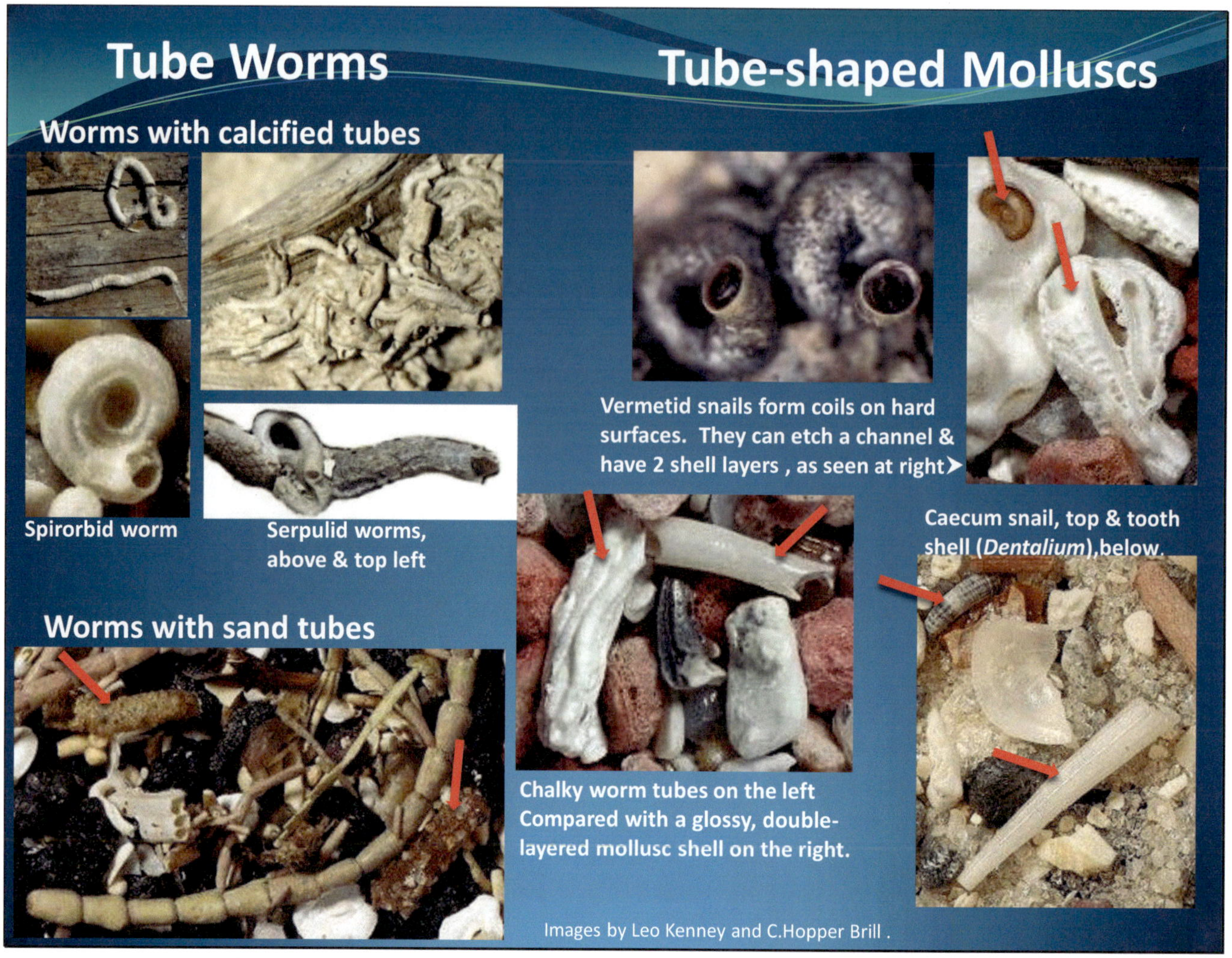

Another pair of animal groups have skeletal remains that look similar: tube worms and tube-shaped molluscs.

Some segmented marine worms (Polychaete annelids) secrete calcite tubes that attach to surfaces ranging from rocks and pilings to algae and the shells of other marine animals. The most common are spirorbid (tight, circular coil) and serpulid (loose or uncoiled) tube worms. Other worm species produce flexible tubes that incorporate sand grains.

Calcified worm tubes can be confused with shells of several kinds of molluscs. Vermetid snails also attach to hard surfaces and uncoil to varying degrees, but their shells possess two layers (outer of calcite and the inner of aragonite). These shells are more dense than the calcite worm tubes and will wear to a polish. Using their rasp-like feeding structure (the radula), vermetids can also etch a channel in the surface they grow on. Other tube-shaped molluscs include *Caecum* micromollusc snails and tooth shells (*Dentalium*, a scaphopod mollusc).

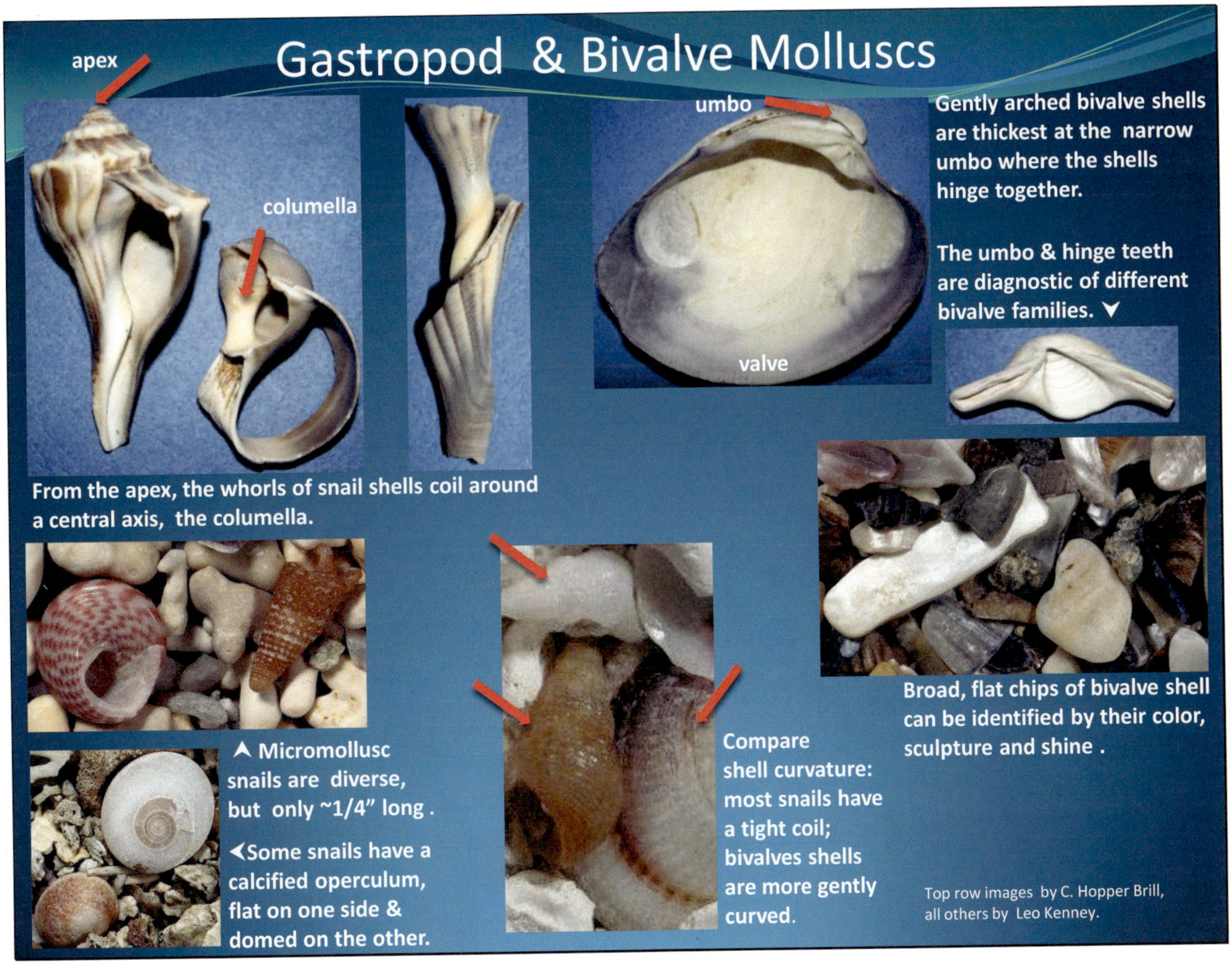

Molluscs have a complex shell with two layers of calcium carbonate, an outer one of columnar calcite crystals and inner layer of flat aragonite crystals. Sometimes, both shell layers can be seen in beach drift. Overall, mollusc shells weather to a porcelain or glassy finish, but with diverse colors and shell sculpture.

Gastropod snails have a single shell that coils around a central column (columella). The top (or apex) and the columella are the most resistant parts of the shell. Snail shells range from domed or volcano shapes to tight cones and spirals; sizes range from a fraction of an inch to several feet in length. Common in sand are micromollusc shells, from snail species no larger than grains of rice.

Bivalves have two shells (valves) held together by an exterior hinge ligament and interior teeth. The broad valves have a much wider arc than most snail shells. The most resistant shell part is usually the umbo which includes the hinge structure. The teeth on the inner surface of the umbo are diagnostic features used in identifying different families of bivalves

While barnacles are crustaceans, relatives of crabs and shrimp, they live attached to surfaces and are encased by a volcano-shaped shell made of several calcium carbonate plates. These plates are common in sands along rocky shores and can be easily confused with fragments of mollusc shell, especially those of bivalves. Here are some clues for telling the two apart:

•In barnacle plates, the wider bases are not solid like mollusc shell, but are honeycombed. Look for parallel openings in the shell.

•Barnacle plates may have a triangular "scar" that looks like the umbo of an oyster, plus rows of grooves that resemble the teeth of some bivalve hinges. But, this combination of scar and grooves isn't found in bivalve shells.

•In contrast to barnacle plates, bivalve shells have 2 layers and wear to a polished surface.

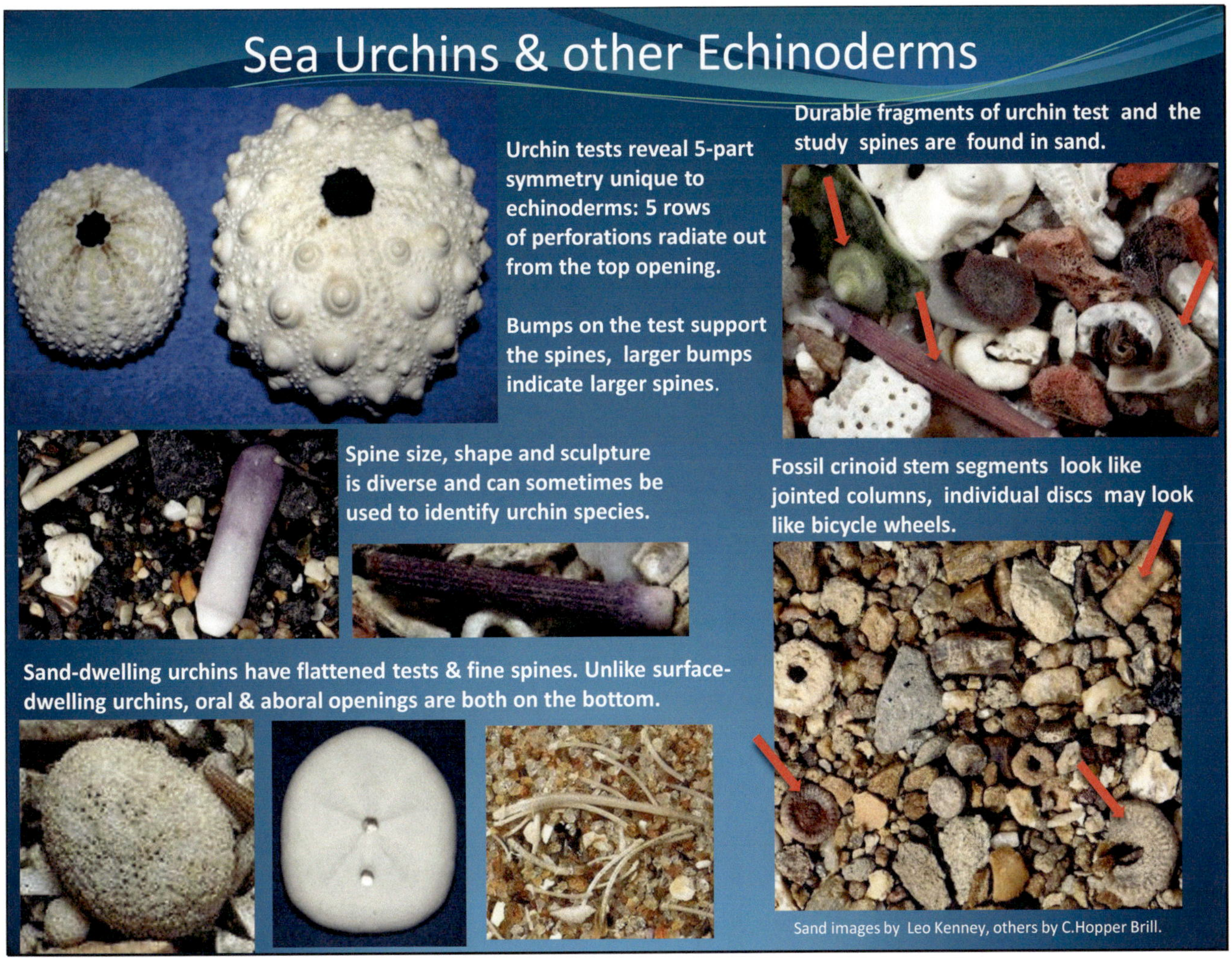

Sea urchins are one of the most recognizable of the Echinodermata, the animal group including seastars, brittlestars, crinoids and other relatives. Keys to urchin identification are: the 5-part symmetry of their internal calcite skeletons (called tests); and their distinctive spines. Look for chalky plates with rows of perforations (for the tube feet) and surface bumps (called tubercles) whre the spines attached. Individual urchin species can be identified based on test structure and spine size, shape and sculpture. Worn spines in sand samples look like grooved columns; in cross section, you see a pattern of concentric rings and radiating spokes.

While many urchins live at the seafloor surface, some – like sand dollars, heart and biscuit urchins - burrow in sand. Their tests are flattened and densely packed spines are very thin. Unlike surface-dwelling urchins, which have oral (mouth) and aboral openings on opposite surfaces of the test, the burrowers have both openings on the bottom surface.

Fossil echinoderms surface in modern sand samples, too. Segments of stalked crinoid (sea lily) stems turn up in both shoreline and inland deposits. Stems look like segmented columns; individual segments may be round, star-shaped or oval in cross-section, their shape and sculpture can be used in identification.

Now, with this primer, what skeletons can you discover in the sand? Biogenous sources add to sands at all latitudes, from the tropics to the poles. Conduct some shoreline forensics by examining sand for clues to the diversity of marine organisms living along the shoreline and below the surface.

Many thanks to Leo Kenney for permitting me to use his wonderful micrographs of sand. Shown here are:
*Top Row, left to right: Beserah, Malaysia, South China Sea; Fanore Beach, Ireland, North Atlantic.
*Bottom Row, left to right: Frangokastello, Crete, Mediterranean; Davao, Samal Island, Davao Gulf, Philippines; Bodrie Beach, Corsica, Mediterranean Sea.

If you have questions, please don't hesitate to contact me by e-mail: chopper@vims.edu.
Carol Hopper Brill, Marine Advisory Program,
Virginia Institute of Marine Science, College of William & Mary,
Gloucester Point, Virginia

Science in a Grain of Sand: Activities for the Classroom

Olga S. Jarrett, Brian Williams, Robert Jarrett, Georgia State University

Mary Gurney, Level Creek Elementary School, Gwinnett County, Georgia

1. **Sand is defined as particles of a particular grain size**, namely 0.064mm to 2 mm. Sand is larger than clay and silt but smaller then pebbles. Quantities of sand can be found many places including beaches, streams and rivers, quarries, coastal dunes, and inland dunes.

This picture taken through a dinoXcope microscope, connected to a computer, shows measurement capabilities.

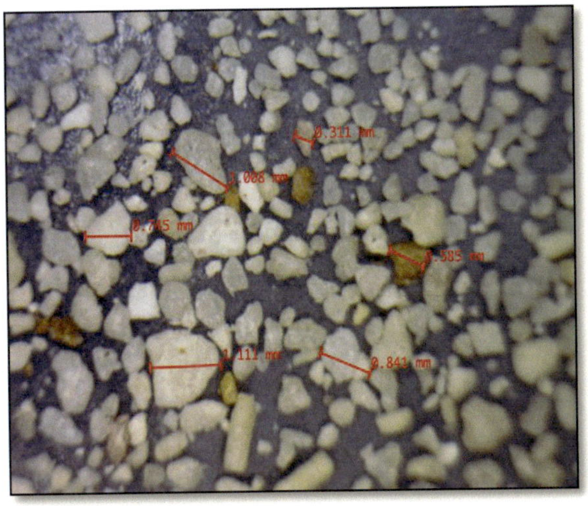

It is possible to measure the size of grains of sand by determining whether they will flow through different size sieves. Commercial sieves can be useful for calculating the weight of sand that will go through each sieve in a series of sieves. The sieve set above has 6 sieves and can allow measurement of grain sizes from 4 mm to .063 mm. These sand sieves were ordered through amazon.com. However, simple household sieves, for a fraction of the cost, can allow calculation of the proportion of grains that will/will not go through a sieve.

2. Sand can be found in soil and helps with drainage. But soils with too much sand drain the rain away and dry out too fast. Experiment with planting seeds in various mixtures of sand and potting soil. Water the pots at the same intervals with the same amount of water. Does the water run out the bottom of the pots faster with one kind of soil? Do the pots with one kind of soil dry out faster that the other? Organic Gardening for Dummies recommends 40% sand, 40% silt, and 20% clay for most plants. Find this and further information at http://www.dummies.com/how-to/content/how-to-assess-soil-composition.html

3. Look at sand with a magnifying glass or microscope. Draw what you see. If the grains of sand are about the same size, the sand is called "well sorted." If they are widely different sizes, it is called "poorly sorted." The picture with the measurements on the previous page is fairly poorly sorted sand. It is sand from a caye (island) off the coast of Belize and is mostly made up of broken or whole corals and shells. In contrast, the following picture through a DinoXcope microscope shows well-sorted sand from Ruby Beach, Olympic National Park, WA.

To view a thin layer of sand under the microscope, make a sand viewer by folding a business card, cutting a wedge from it and taping a piece of tape on the back across the hole. Press the sticky side in sand and pat it down. Then you can look at it under the microscope.

This sample from Hannahatchee Creek, Stewart County, GA has tiny fossils in it. Look at how sharp or rounded the grains are in your sand sample. Rounded grains have been rubbed against each other by wind or water. Waves tend to round sand, so sand from beaches with much wave action is likely to be more rounded.

The sample to the right is from the Chesapeake Bay, and like the sand from Ruby Beach, it is well sorted and fairly rounded. It is made up of quartz and probably other minerals.

4. Put sand in a little vinegar or add a few drops of vinegar to a small pile of sand. Does it bubble? If so, the sand is probably biogenic, made of coral and shell pieces, calcium carbonate ($CaCO_3$).

If it does not bubble, it is probably quartz sand. The picture to the left shows the calcium acetate, $Ca(CH_3COO)_2$ that is left in a milk bottle cap after the chemical reaction between the vinegar and biogenic sand from Belize.

5. If you see dark specks in sand, there may be special mineral particles. Test with a powerful magnet in a plastic bag and see whether you can drag the particles toward the magnet. If so, you have bits of magnetite. (The bag keeps the particles from sticking almost permanently to the magnet.) In the picture to the right, sand from Costa Rica was in a jar, and a rare earth magnet was placed under the jar. This sand, from the southern Caribbean coast of Costa Rica, is made up largely of magnetite.

6. Sand ripples. Air and water can cause sand to ripple like tiny dunes. The ripples below are from Jekyll Island, GA and are caused by the wind.

The ripples below, also from Jekyll Island are caused by the tides.

The following idea for creating ripples is adapted from the website: http://sand.xboltz.net/ripple_marks.pdf.

Place a glass in the center of a pan (I used a disposable plastic cake lid) and add washed sand and water. Stir the water in one direction without digging into the sand. Watch the ripples form in the sand from the eddies that are created by water movement around the glass.

7. **Contact schools** in other parts of the US and the world and ask them to send you sand. Offer to trade with them. Compare these sands with the sand in your area. Your class may develop pen pals through sand collecting.

8. **Sand dunes** are created when the wind causes sand to pile up. Dunes are found in many places, including beaches and inland areas where there were former beaches or where the wind conditions are such that sand collects in one area and can't escape. One such place is Great Sand Dunes National Park in Colorado.

The sand in these dunes comes from the surrounding mountains. The sand is blown back and forth within a valley. The following Google Earth image shows the dunes on the west with the mountains that prevent the dunes from moving on the east.

Sand in dunes is mounded up but can get only so steep and then the sand runs down the sides like an avalanche. The angle at which dry sand runs off is called the angle of repose. The angle of repose for dry sand is 34°.

Try making a pile of sand by pouring sand through a funnel onto a paper plate. Line up a protractor behind the pile with the straight bottom centered at the top of the dune.

Do you get 34°? The picture to the right taken in a quarry shows a pile of sand or gravel. Can you estimate the angle of repose of the pile in the picture?

You can find a list of various materials and their angle of repose at http://en.wikipedia.org/wiki/Angle_of_repose

9. Set up sand learning stations in your classroom. In the following station, students can make their own sand in a coffee can by shaking stones in the can, with or without water. When finished, allow the can to sit briefly, then open the lid away from you so you do not breathe sand dust.

The picture below is of a learning station in which students are to match sand samples from various parts of Georgia with their location on the map. Sand can be found in creeks, rivers, and dunes in many parts of Georgia. Mountain sand is closest to its source and has the poorest rounding and sorting.

Coastal sand originally came from the Appalachian Mountains but underwent a lot of grinding and polishing on its way to the coast. Beach sand is likely to have very small particles and be well sorted and somewhat rounded.

The microscope is a cordless 20X and 40X Cynmar stereo microscope with above and substage illumination. It can be recharged and operates many hours per charge. Check on line.

10. Sand art. The Navaho include native symbols in sand paintings.

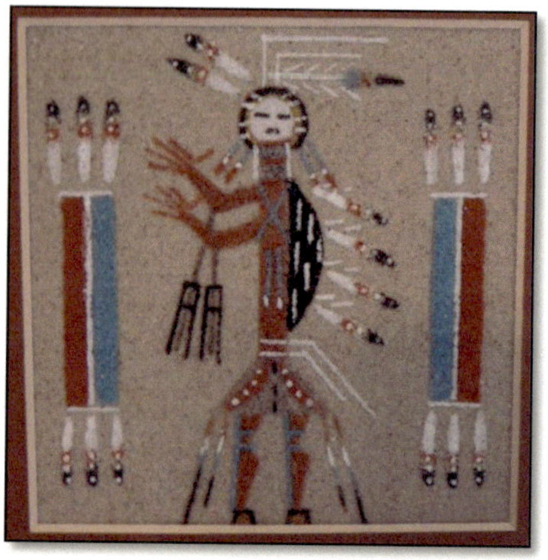

Tibetan Buddhist monks create Mandalas, elaborate designs with artificially colored sand.

These designs are for religious purposes and are destroyed, with the sand poured into a creek or river, at the end of the ceremony.

Sand naturally has many different colors including almost white, tan, greenish, reddish, yellowish, pink, and black. You can make your own sand art with naturally colored sand. Work on a small piece of plywood, cardstock, or cardboard. Draw a design. Fill in where you want one color with white glue using a paintbrush or Q-tip. Sprinkle sand on the glue. Then follow with more glue and other colors until you fill in the whole picture. This sand painting was made from natural sands of Providence Canyon, GA.

Sand Castle Recipes

Professional Sand Castle Stuff

Ingredients

10 parts sand by volume

1 part white glue by volume

3 parts water by volume

No more washed away sand castles! This mixture of sand and glue to help hold together your work of art. Mix sand, water and glue together in a large bucket. If the mixture is too thick, add more water. Expect your sand sculpture to last weeks — even if it rains!
http://preparednessnibblesandbits.blogspot.com/2009/04/professional-sand-castle-stuff.html

Indoor Sandcastle Stuff

Ingredients
2 cups clean, fine sand
1 cup cornstarch
1 cup water
Bring the beach inside — even in the winter! The cornstarch in this recipe holds the sand together. Mix all three ingredients over a low heat (200°F). Remove when the mixture firms. This mixture can be carved with spatulas and table knives.
http://preparednessnibblesandbits.blogspot.com/2009/04/indoor-sand-castle-stuff.html

A Mock Rock Recipe (I have tried this one)

1 cup flour
1/2 cup salt
2 teaspoons alum
1 cup coarse sand
1/2 cup water
food color (5 drops red,
3 drops blue, 3 drops yellow)
mystery rocks (rocks to find)

First, mix the dry ingredients in a bowl. Add food coloring to the water. Pour some water into the dry ingredients and knead the "rock dough." Slowly add more water until it feels like play dough. Then, wrap pieces of the dough around the mystery rocks. Set on cookie pans to dry. Let the rocks dry. It takes about a week. Turn the rocks everyday or two so they dry evenly. Allow children to excavate.
http://saltthesandbox.org/rocks/mockrock.htm

Bibliography

Greenberg, G. (2008). *A grain of sand: Nature's secret wonder*. Minneapolis, MN: Voyageur Press. [Amazing micro-photographs of sand from all over the world.]

Prage, E. J. (2000). *Sand*. Washington, DC: National Geographic Society. [Very readable book by a geologist on the origin and nature of sand.]

Steinmetz, G. (2012, November). Sailing the dunes. *National Geographic*, 94-109

http://waynesword.palomar.edu/ww0704.htm#Introduction (about dunes)

Welland, M. (2009). *Sand: The never-ending story*. Berkeley, CA: University of California Press. [Excellent children's book on various types of sand, what sand looks like close up and how sand is created]

What do you know about Sand?

Here are three activities that Mary Gurney has used with elementary school students for problem solving and to assess understanding.

Activity 1: Which One is Which?

Materials: Small containers of a variety of sands with a number and information on the bottom of the small containers, numbered cards with the same variety of sands stuck to them, magnifiers, microscopes [if available], millimeter measures, and magnets.

Procedures: Use all your observation skills to try to match each card to the correct container of sand without turning the container over. You may use the science tools. When you are done, see if the number on the card matches the number on the bottom of the container. Make a list of the types of clues and tools that you used. Write about which clues and tools helped you the most.

Activity 2: The Tools Help Find Answers

Materials: Magnifiers, microscopes, millimeter measures, magnets, sand sample cards, and trivia board with questions on a flap and the answer underneath.

Procedures: Use the science tools and numbered sample cards to help answer the trivia questions on the trivia board. Write the number on each sample card next to the question it matches. Flip up the trivia question to see the numbers of the correct sample cards. Write a sentence about what surprised or interested you the most.

Activity 3: Sand and Rock Relationships

Materials: Pairs of rocks that can be rubbed together (e.g., granite, gneiss, sandstone, marble, limestone, and soapstone), paper to "catch" rubbed particles, magnifier, microscope, and sand sample cards.

Procedures: Pick up two matching rock samples. Rub them together firmly over the paper to collect the "weathered" sample you create. Look at the "weathered" particles with the magnifier or microscope. Rub the particles between your fingers. Try another pair of rock samples.

> Which rock samples created sand?
> Which did not create sand?
> Write a sentence or two to explain the difference between the two samples.

Now look at the sand sample card provided. See if you can determine which minerals could be part of the "parent" rock before it weathered to this sand. Turn over this instruction card to find the answers on the back.

Response Page to Sand Activities

1. **Which One is Which?**

Which types of clues and tools did you use?

____

____

Which clues helped you the most?

____

____

2. **The Tools Help Find Answers**

Question	Matching Sand Card Sample Number
Which sands are MOSTLY quartz?	
Which have iron pyrite?	
Which sand is made of as much or more shell than sand?	
Which sand was from inside a geode?	
Which are attracted to a magnetic?	

3. **Sand and Rock Relationships**

Which rock sample created sand?

____

Which rock sample did not create sand?

____

Explain the difference between the two samples.

____

____

Deborah Wills

The Sand Beneath My Feet

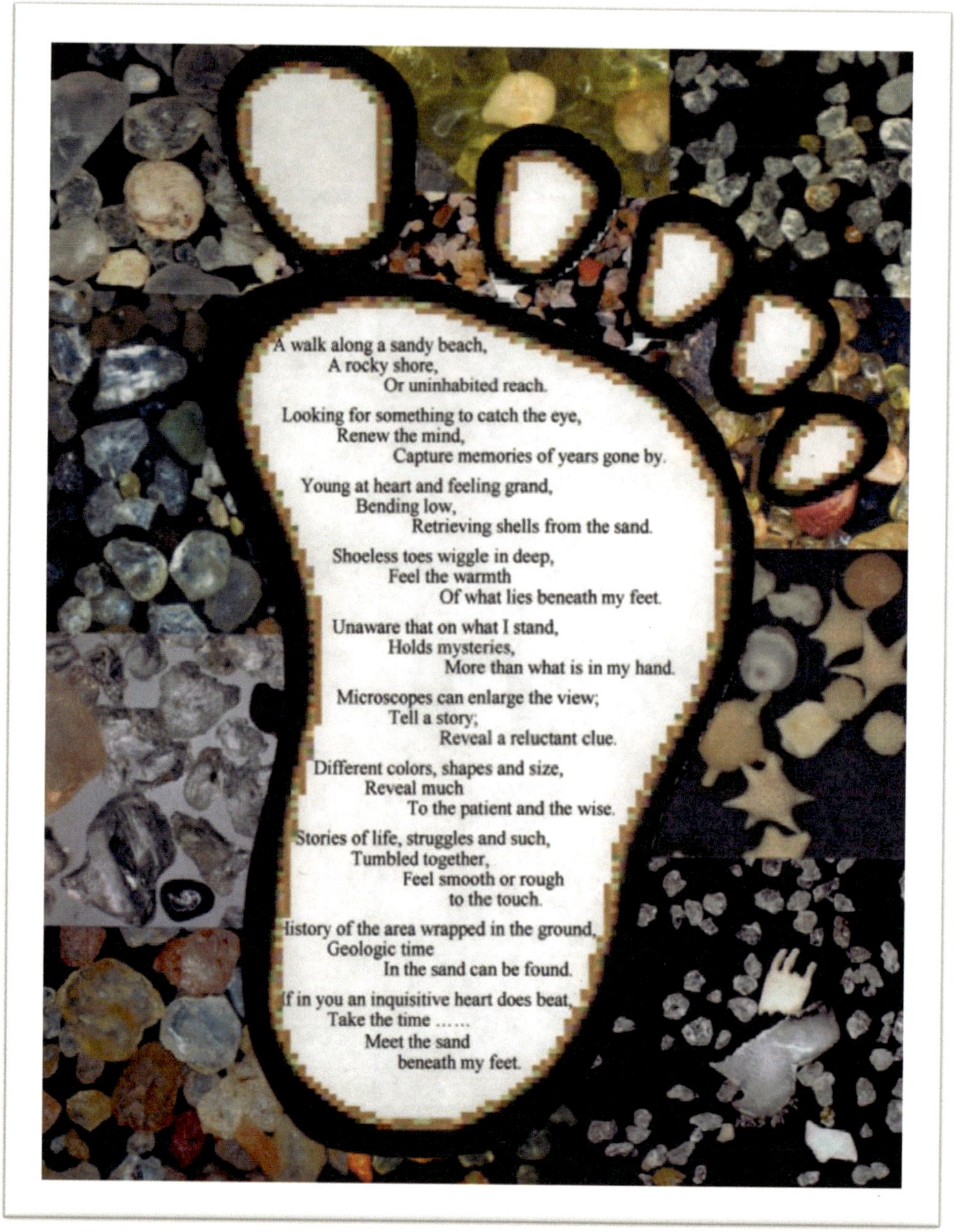

Sand (clockwise from top left): Key West, FL; Green Sand Beach, HI (1); Pismo Beach, CA; Botany Bay, SC; Green Sand Beach, HI (2); Star Sand – Pacific Ocean; Mobile, AL; France; Decatur, AL (TN River); Tybee Island, GA; San Diego, CA

Heidi Scheirer McGrew

Kanuka Grove No.3
(11″ x 14″)

Sunset, Anna Maria Island
(11″ x 14″)

Detente
(8″ x 10″)

Into the Garden
(5″ x 7″)

Snowy Trail, Dolly Sods
(9″ x 12″)

Running Beach Eve No. 2
(8″ x 12″)

Barn Owl Circle No. 3

Otama Cliffs No. 1
(8″ x 10″)

Jose Santamaria

Sand Tales
Jose Santamaria
1871...Stanley finds Livingstone...
sinking in quicksand
2013
"Dr. Livingstone, I presume you need
more little bags?"

Sand Tales
Jose Santamaria
2013
1916...sand collector Lawrence of Arabia
finds his perfect grain

Sand Tales
Jose Santamaria
Yet another excuse for being late to work
2013
"Sorry boss somebody stole all my sand!"

Lori Carter

Sand Collector's Nightmare

Find anything good?

Okay, sandman - check! Next - quicksand!
sands of the world
sandstone
sandwich
ampersand
&
sand castle
sand timer
sandman
sand

Made in the USA
Las Vegas, NV
04 November 2026

57424977R10155